VIRTUAL TESTING OF MECHANICAL SYSTEMS

ADVANCES IN ENGINEERING

Series Editors:

Fai Ma, Department of Mechanical Engineering,
University of California, Berkeley, USA

Edwin Kreuzer, Department of Mechanics and Ocean Engineering,
Technical University Hamburg-Harburg, Hamburg, Germany

VIRTUAL TESTING OF MECHANICAL SYSTEMS

THEORIES AND TECHNIQUES

OLE IVAR SIVERTSEN

**Norwegian University of Science and Technology (NTNU)
Trondheim, Norway**

Psychology Press
Taylor & Francis Group

New York London

Library of Congress Cataloging-in-Publication Data

Sivertsen, Ole Ivar, 1945-
Virtual testing of mechanical systems : theories and techniques/ Ole Ivar Sivertsen.
p. cm. -- (Advances in engineering ; 4)
ISBN 9026518110
1. Computer simulation. I. Title. II. Advances in engineering (Lisse, Netherlands) ; 4.

QA76.9.C65 S567 2001
003'.3--dc21

2001034390

First Published by Lawrence Erlbaum Associates, Inc., Publishers
10 Industrial Avenue
Mahwah, New Jersey 07430

Reprinted 2010 by Psychology Press

Cover design: Ivar Hamelink
Printed in the Netherlands by Grafisch Produktiebedrijf Gorter b.v., Steenwijk

ISBN 90 265 1811 0

Contents

Preface

The purpose of writing this book is to give users and potential users of the multidisciplinary simulation code FEDEM an understanding of some of the basic theories and techniques used in this simulation program. Parts of this text have been used for some years in a course that I have been teaching for first year graduate students. Some of the mathematical fundamentals for the theory presented are included in the book to make the text self-contained also for undergraduate students who do not have the proper mathematical background. Some basic finite element techniques are also included. However, to get a fundamental understanding of the method presented it is recommended to have some previous knowledge of finite elements before reading the text.

Because of the multidisciplinary nature of this simulation code and the aim of the book to present both theories and techniques, the style and to some extent the notation are somewhat different from section to section. Examples from the wide span of subjects covered by the book are product design theory, matrix mathematics, linear and nonlinear finite element theories and techniques, kinematic constraint modeling, details of transmissions, integration methods and last, but not least, control engineering.

Demanding mathematical formulations and proofs that, in my opinion, do not support the overall understanding of the simulation flow or the physical effects that are to be modeled, are excluded from the presentation. In other words, this is not the text to read if the main purpose is to study nonlinear finite element formulations, classic mechanism theory or control engineering theory in general. However, if you like to see how theories from different fields of engineering are applied in a multidisciplinary manner this text may be something for you. The book is quite self-contained regarding the basic mathematical techniques required for reading the text.

Chapter 1 presents some basic knowledge about large computer simulation, gives some simulation definitions and puts simulation in the framework of modern product design philosophy. The chapter concludes with some industrial simulation applications.

Chapter 2 present fundamentals such as matrix and vector mathematics, geometric transformations, finite element techniques and some basic numerical algorithms.

Chapter 3 covers the structural and kinematic modeling aspects of mechanisms such as the mechanism design position, substructures and super elements, updating of super elements, geometric stiffness, master-slave constraint and transmission modeling, spring constrained joints and transmissions. Other mechanism elements including extra inertia, gravitation, prescribed motion and initial condition are also covered.

Chapter 4 covers the formulation of the structural dynamic equation of motion including the formulation of the integration algorithm. The control engineering part is formulated as a separate unit describing the interaction with the structural part both with respect to data transfer and integration algorithm interaction. An overview of primary and secondary simulation variables including control, stress retracking and the eigenvalue results are presented in one section. The Chapter concludes with algorithms for energy calculations.

Chapter 5 gives an introduction to how the computer program based on the formulations and techniques presented in this book can be used as a design tool. First a general description is given, then a simple design case is taken through an interactive design optimization process visualizing typical operation using the FEDEM code.

The text in the appendices is either based on mathematical theory not presented in this book or judged to be of a nature that is not general enough for most potential readers of the book.

The multidisciplinary compilation of theories and techniques in this book includes major contribution from several colleagues at the Norwegian University of Science and Technology (NTNU) or research/industrial organizations here in Trondheim, Norway. Sections 1.2.2 and 3.6.3, covering design philosophy and detailed modeling of transmissions, respectively, are to a large extent contributed by Hans Petter Hildre, professor in mechanical engineering, NTNU. Section 4.2 covering control in mechanism simulation is contributed by Dr. Torleif Iversen, senior research scientist in control engineering, SINTEF.

Fedem Technology employees, Dr. Bjørn Haugen, Siv.ing. Dag Rune Christensen and Karl Erik Thoressen have contributed by developing and formulating Section 4.4 Energy Calculations, Appendix A Corotational Geometric Stiffness and Appendix B The Cam Joint, respectively. The interactive graphic interface to the simulation program, referred to in Chapter 5, is to a large extent the work of Siv.ing. Jens Lien, Jens Jacob Støren and Geir Moholdt. Terje Rølvåg, also a Fedem Technology employee, has made contributions to most of the R&D work done developing the FEDEM software over the last 12-14 years and has contributed specifically with the applications in Section 1.4.

Last but not least I will acknowledge Kolbein Bell, professor in civil engineering, for his well structured package of versatile and dependable algorithms and techniques for modeling of general finite element based simulation codes, referred to in Section 2.3.

On the non-technical side, I acknowledge the editorial assistance from cand. philol. Stewart Clark, NTNU.

I am especially grateful to these people, as well as others, not mentioned here, for their contribution to the theories and formulation in this book.

Ole Ivar Sivertsen

Chapter 1

Introduction

1.1 Multidiscipinary Simulation

1.1.1 What is Multidisciplinary Simulation?

Multidisciplinary Simulation(MDS) is a method of electronic data processing that permits a product to be simulated, i.e. to be calculated as a complete system or a composition of its different components on a computer. A virtual prototype of the product is being built into the computer, not only for visualizing the details of the design, but also for studying the physical behavior of the proposed product with respect to functionality and safety. The numerical results can be visualized by the means of diagrams, single pictures and animated motion sequences on a computer screen.

1.1.2 How is MDS Used?

MDS is mainly used during the product development process. As MDS is by definition completely independent of the existence of a real product, the engineer is enabled for instance during the development of a product - without the existence of any product in reality - to identify the characteristics of the product on a computer, to simulate, to calculate it and to visualize it on a computer screen. The development engineer can calculate all kinds of product reactions for instance the behavior of a vehicle when cornering, when going over bumps, etc., how the vehicle reacts, when the drivers initiate certain steering manipulations, when they accelerate, when they brake, etc. The development engineer can investigate, which requirements concerning behavior, fatigue, strength, dynamic characteristics a certain product component has to fulfill, so that the product has a desired overall behavior. These theoretical investigations can be performed at a very early stage a long time before experimental investigations with a prototype of the product will be possible.

1.1.3 How does the Customer Benefit from the Use of MDS?

The product development can be accelerated by means of MDS processes. This means that the customers for the product will have the advantage of being faster served at reduced cost but they are nevertheless provided with optimum results. In other words a high-quality product is available to the customer much earlier than at present. By using MDS on the computer, vehicle optimization regarding comfort and safety can be performed a long time before any prototype will be built. Cost-intensive test work can be reduced in this way. If these investigations had been done just by tests or by experiments, this would mean time- and cost-disadvantages for the customer. Another point to take into account is, that by means of system simulation many new product ideas can be checked and a large number of varieties of technical solutions can be proven. Also ideas and varieties can be simulated that cannot be investigated by tests or by experiments because this is very time consuming and requires too much effort. In this way the user of MDS can offer the customer highly qualified solutions, that would not have been possible without MDS, solutions which place one in a better position in comparison with the competitors.

1.1.4 Which Electronic Data Processing Methods are Available for MDS?

On the one hand, there is the so-called full product simulation which enables the overall behavior of the total product - as perceived by the user - to be simulated, i.e. calculated on the computer. This is also called *Multi Body Simulation* (MBS), where the real physical behavior of the product under investigation can be reduced to a few characteristics, for instance overall behavior, and then simulated as a numerical product model on the computer.

On the other hand, the *Finite Element* (FE) method is used for instance for a vehicle to calculate fatigue, stiffness, dynamic behavior of the car body, of chassis components, of engines. The FE method is also used for calculating the temperature distribution, it is used to investigate gas dynamics in the combustion chamber of an engine, etc. When applying this method of FE on a product part, the component under calculation, for instance the car body, is numerically broken down into many small parts on the computer; these individual parts are calculated separately and the results are then reassembled again. The finite elements size and shape are chosen according to the required accuracy of the results. For many years the finite element method has been a successful tool for the solution of product analyses with respect to functionality and safety.

Multidisciplinary simulation combines multi body simulation, the finite element method and *control engineering*. These computer programs do not replace the traditional finite element programs, however, they make much more detailed modeling of the overall system possible and consequently much more accurate simulation results.

These types of investigations that require a lot of computing time for the simulation can only be performed in many cases with high-performance computers. Nowadays this means for instance most UNIX-workstations, advanced PCs or supercomputers like the CRAY.

1.2 Some Definitions

1.2.1 Simulation Definitions

Simulation

By *simulation* we mean the imitation of certain properties of a system in a mathematical computational model.

Model

A *model* in this context is defined as a more or less simplified representation of the engineering aspects of a system. The model must represent the properties of the aspects of a system that you would like to study, as accurately as possible.

Mechanism

A *mechanism* is a device which transforms motion to a desirable pattern. It is an assembly of links where at least two of the links have relative motion. The links may be connected to each other by joints that constrain their relative motion.

Machine

A *machine* typically contains mechanisms in addition to energy transmission functions. Single parts in a machine are called machine elements that are connected to each other in an assembly. A link in a mechanism can be built from several machine parts and be in one or several assemblies.

Kinematics

Kinematics refers to calculations of motion in a system with no reference to forces and torques in the system or the input necessary to achieve the motion.

Kinetics

Kinetics sets the motion in relation to the forces and torques. Kinematics and kinetics are not physically separable. We separate them for theoretical reasons. It is also valid in engineering design practice to first consider the desired kinematic motions and subsequently investigate the kinetic forces associated with these motions.

Dynamic Simulation

Dynamic simulation refers to the calculations of motion in a mechanical system on the computer where both constraint forces and the forces necessary to drive the system are taken into account. Also refer to the definition of kinetics and kinematics above.

1.2.2 Simulation in Product Design

Aim of the Simulation

The aim of the simulations are:

- to understand the behavior of the proposed product
- to predict the dynamic response of the product
- to predict the behavior of the product
- to predict the feasibility regarding safety requirements

Simulation as a Design Tool for the Engineer

- to appraise and compare different design concepts
- to improve system properties
- to control dynamic responses

- to reduce production costs
- to be used in dimensioning
- to get an understanding of which parameters affect the desired performance most (parameter sensitivity)
- to be used in optimalization
- to reduce prototyping
- to reduce product development times

Product Development of Machine Systems

The *product development* is the process of transforming the product requirement specifications into product specifications.

The Product Requirement Specification

The requirements that the product must fulfill could be typically:

- functionality requirements
- behavior requirements
- boundary conditions

The Product Specification

The *product specification* is usually available as verbal descriptions and machine drawings of individual parts and the whole product assembly, typically including:

- external form
- materials
- standard parts
- dimensions and tolerances
- surface finishing
- production specifications
- assembling procedures

The Conventional Product Development Process

The product development of high-performance machine systems is usually a multidisciplinary and complex task. Typically the development process are divided in three main activities:

1. Selection of concept and kinematic topology, where the latter includes:
    - decisions regarding number of joints
    - selection of joint types
    - selection of link lengths
    - selection of joint orientations

   These data are derived from the functionality specified in the requirement specifications for the product. The required number of rigid-body degrees of freedoms and the available working space for the mechanism are typical constraints. Mobility analysis is used to calculate the number of rigid-body degrees of freedom the selected mechanical system will obtain. Mechanism synthesis and kinematic analysis may at this stage be used for the selection of link lengths to satisfy the requirements for geometric positions and motions.

2. Design and selection of components to transform the selected concept into a product. This transformation will include the selection of geometric form, materials and production processes. Complex products are broken down into subsystems with their own specifications from the product requirements. At the lowest subsystem level the following activities may be identified:
    - identification of components
    - selection of materials and production processes
    - identify function, boundary conditions and flow (force, heat, etc.)
    - design of contact surfaces
    - component design
    - integrate more functions to each component by reorganizing the design

3. Dynamic simulation and optimization are conducted to evaluate the responses of forces and torques and the overall behavior of the product and its subsystems. The simulations should be used iteratively to update the design towards an optimum with respect to customer

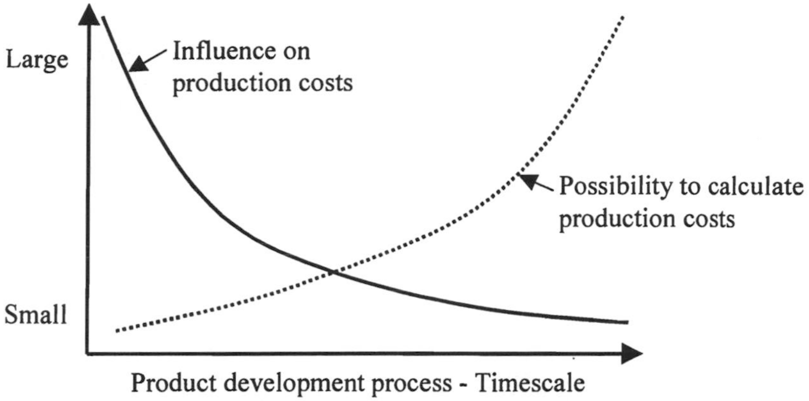

Figure 1.1: Influence on production costs and possibility to calculate production costs as a function of the product development process.

satisfaction through improved technical functionality and reduced production costs.

Concurrent Design versus Conventional Design

It is during the conceptual design phase that the designer can influence the properties of the final product most, and consequently the total production costs, see Fig. 1.1. The dynamic properties and responses for a machine system is usually essential, therefore dynamic simulation has to be utilized as a tool as early as possible in the design process. Doing the different product development tasks in a sequential order is inefficient and time consuming. *Concurrent design* means that all product development are being done simultaneously both the tasks for developing the product and planning the production process for the same product. The development work is conducted in a team of people covering the different disciplines of product development from conceptual design to production processes. The disadvantages with a conventional design process are typically:

- it is difficult to accomplish overall optimization of the product because the development work is distributed between different departments (suboptimalization)

- a set of independent analysis tools utilized by different design teams increases the manual work and the probability for introducing model errors

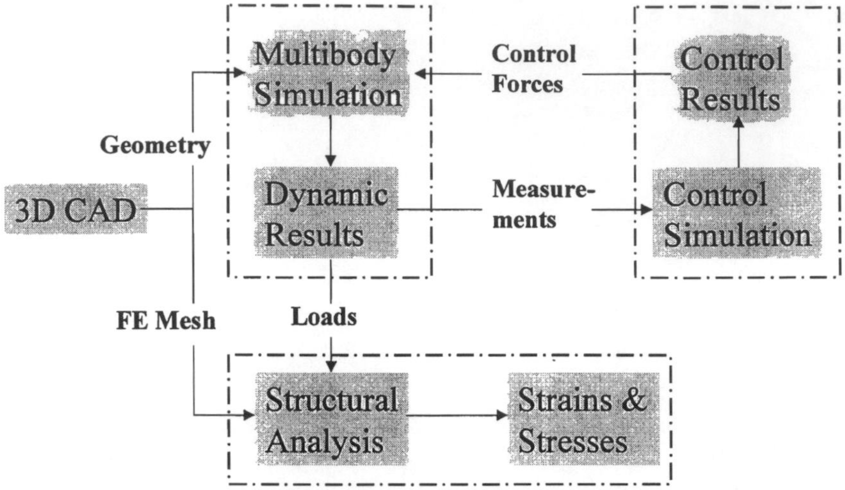

Figure 1.2: Technological islands.

- sensitivity analysis and numerical optimization are not available at system level for the designer. Sensitivities are used to decide which design parameters will influence certain product properties the most

A concurrent design approach increases the creativity and are more efficient.

1.3 The New Simulation Philosophy

In Fig. 1.2 we can see a typical situation for product simulation in industry. Separate teams, and in many cases even separate departments, are working with Multibody simulation, Control simulation and Structural analysis, respectively, see the boxes in the dashed-dotted lines. Communication between these boxes is usually data conversions between stand-alone programs that are controlled more or less manually. Besides this, the engineers in these different teams are specialists in their own fields and often have little knowledge about the other fields. This transfer of data and information are often referred to as "over the wall" situations both mentally and physically. This could be very harmful for achieving a true concurrent engineering design process. The simulation activities are done sequentially that is the simulation results from one team is the input to the simulations by the other teams, and so on, and the consistency of the simulation results cannot be guaranteed.

Fig. 1.3 shows how the new simulation philosophy simplifies the picture and how true integration of the simulation disciplines is accomplished

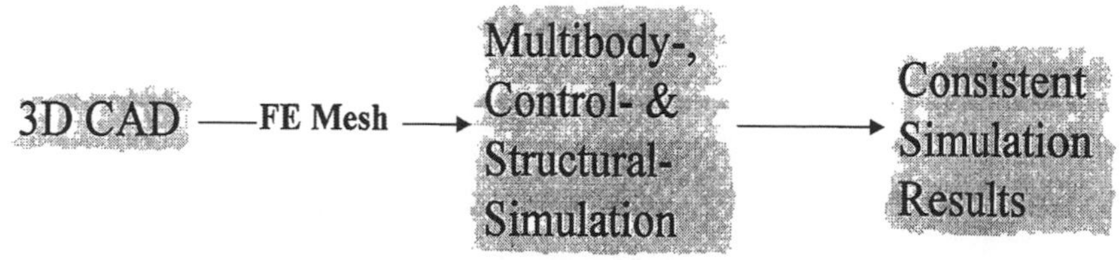

Figure 1.3: Bridging technological islands.

through this new software structure. In addition, this also takes care of the coupling effects between the simulations. As design is an iterative process, the first simulation models are usually rather simple. The objective is to increase the designers' knowledge and investigate possibilities. During the design process the simulation models will be made more and more accurate. Seamless integration is needed to encourage repeated iterations towards the optimal design. This is unlike the conventional approach where a lot of time and effort is needed to update and exchange files and administrate data.

Good communication and concurrency of the work by the different specialists are a must in this situation and this new simulation philosophy will have consequences for the way simulation teams are put together. To fully utilize the potential of this software, simulation should be used as a design tool throughout the design process and not only for verifying a completed design as often is the case.

1.4 Multidisciplinary Simulation Applications

1.4.1 Floating Crane

Fig. 1.4 shows the simulation model for a lifting operation on a FPSO (Floating Production, Storage and Offloading vessel). This is much more demanding for the crane and the crane driver than similar operations on a fixed offshore platform. Due to the motion of the FPSO in heavy sea and strong wind, the crane is subjected to additional dynamic forces as well as swinging loads.

A paper has dealt with numerical simulation of the dynamic behavior of the Norne FPSO offshore crane during lifting operation, see Langen, Birkeland and Rølvåg (2000). The simulations were performed by FEDEM - a general nonlinear dynamic analysis program for flexible multibody systems. The pedestal, "king" and boom are flexible links modeled by shell finite elements and connected together by different joints. The hoisting rope and the

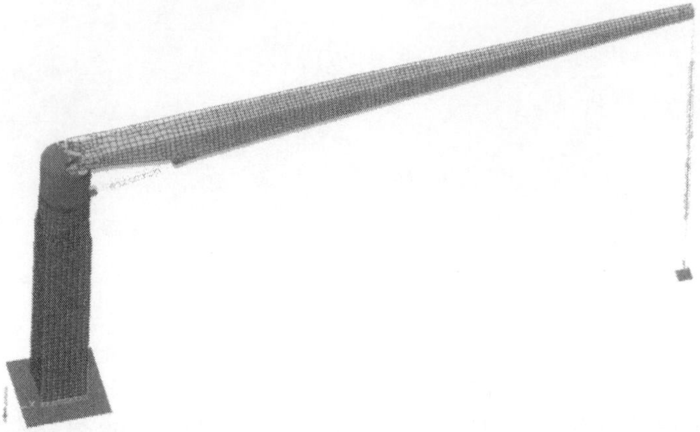

Figure 1.4: Simulation model for floating crane.

hydraulic cylinders are modeled by linear and nonlinear spring-and-damper elements. A control system is implemented in the model making it possible to control the movement of the boom and the winch to compensate for the relative motion between the ship and the supply vessel and keep the load at rest relative to the vessel. Examples are shown of calculated natural frequencies and mode shapes as a function of hoisting rope length and boom angle. Furthermore, maximum dynamic stresses in different sections/details are presented as a function of how the crane is operated.

Besides giving necessary stress data for design verification against overload and fatigue, the presented model can be used to optimize the operation procedure, determine the maximum allowable load for various sea states and to calculate necessary power to control the motion of the load.

1.4.2 Gomos Satellite

Fig. 1.5 shows the simulation model for the GOMOS (Global Ozone Monitoring by Occultation of Stars) instrument developed by Matra Marconi Space for the low Earth orbit ENVISAT-1 mission of the European Space Agency. The purpose of GOMOS is to provide a daily global geographical coverage of the vertical distribution of ozone by stellar occultation technique.

The GOMOS Steering Front Assembly (SFA) is essentially a twin-axis pointing system. It comprises of a mirror (SMU), a Mechanism Drive Electronics unit (MDE) and a Steering Front Mechanism (SFM). Unlike many other pointing systems the SFA must maintain a line of sight stability better than 3 microrads, while subjected to considerable microvibration distur-

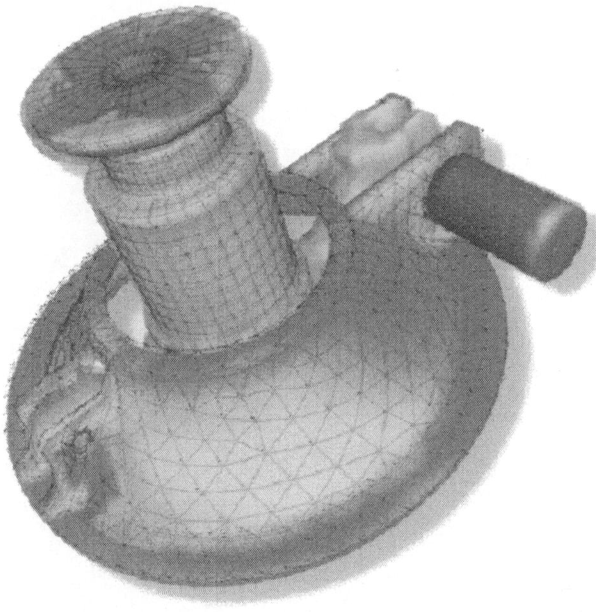

Figure 1.5: Simulation model for Gomos satellite.

bances from the host spacecraft. A paper describes the key design drivers, the mechanism architecture, and how FEDEM simulation was used to identify and solve the problems that were found during testing of the first GOMOS prototype, see Rølvåg and Humphries (1998).

1.4.3 Moxy Truck

Fig. 1.6 shows the simulation model for the MOXY Dumper. Moxy Trucks AS is a company producing a line of advanced and reliable Articulated Dump Trucks (ADT). Moxy's creative efforts in CAE-based product development have lead to many technological innovations and their new models have several advantages. The soft outlines of the design along with raised standards in comfort are exceptional when compared to traditional construction machinery.

By the use of FEDEM software and services, the new generation of Moxy ADTs have been refined and developed further to meet future demands.

Figure 1.6: Simulation model for Moxy Truck.

1.4.4 Neos Robot

Fig. 1.7 shows the simulation model for the Tricept 805 robot from Neos Robotics AB that is a second-generation parallel-kinematics machine specifically designed for milling applications. Parallel-kinematics machines, sometimes called "machine tool-robots," show great promise because they combine the strength of a linear machine tool with the speed, flexibility, and precision of a robot.

Neos has used FEDEM software and services for structural testing and optimization of all Tricpet machines to ensure that the technology is viable. Physical testing has proven that the simulation results have been very reliable. All eigenfrequencies have been predicted with an accuracy of 5% or better in all Tricept configurations.

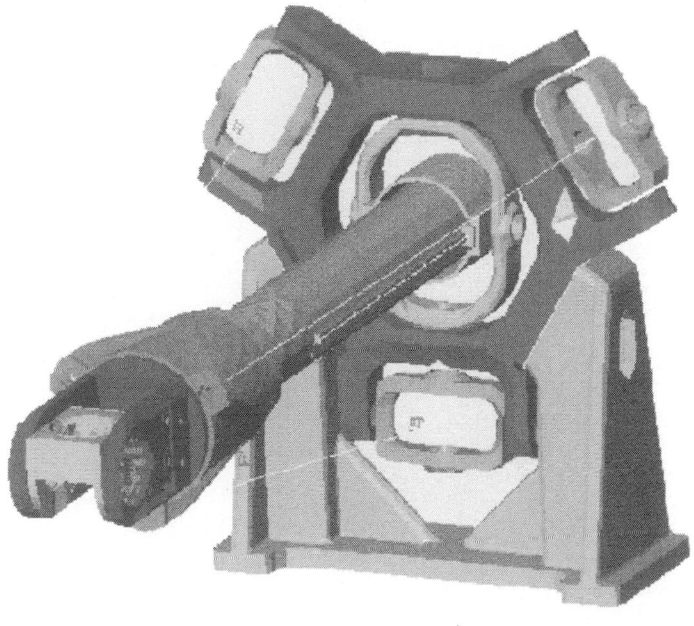

Figure 1.7: Simulation model for Neos robot.

Chapter 2

Theoretical Fundamentals

2.1 Matrices and Vectors

2.1.1 Definitions and Notations

A *matrix* is defined as a rectangular array of numbers arranged in m rows and n columns, and the dimension of such a matrix is written $m \times n$. In the general case, m and n are greater than 1 and a matrix will be referred to in one of the following ways

$$\boldsymbol{A} = \{a_{ij}\} = \begin{bmatrix} a_{11} & a_{12} & \dots & a_{1n} \\ a_{21} & a_{22} & \dots & a_{2n} \\ . & . & \dots & . \\ a_{m1} & a_{m2} & \dots & a_{mn} \end{bmatrix} \tag{2.1}$$

where element a_{ij} is located in the ith row and j^{th} column. *Vector* in matrix algebra refers to a matrix with only one row or only one column. For example, the following represents a column vector of dimension $m \times 1$.

$$\boldsymbol{a} = \begin{bmatrix} a_1 \\ a_2 \\ . \\ . \\ a_m \end{bmatrix} \tag{2.2}$$

2.1.2 Matrix Operations

Addition

Having two matrices $\boldsymbol{A}$ and $\boldsymbol{B}$ with the same number of rows and columns, they may be added to produce a resultant matrix $\boldsymbol{C}$ whose elements are equal to the sum of the corresponding elements of $\boldsymbol{A}$ and $\boldsymbol{B}$; i.e.,

$$\{a_{ij}\} + \{b_{ij}\} = \{c_{ij}\} \tag{2.3}$$

where

$$a_{ij} + b_{ij} = c_{ij} \tag{2.4}$$

Example 2.1 : Matrix addition.

$$\begin{bmatrix} 1 & 5 & 2 \\ -6 & 3 & 7 \end{bmatrix} + \begin{bmatrix} 0 & 2 & 8 \\ 6 & 15 & -4 \end{bmatrix} = \begin{bmatrix} 1 & 7 & 10 \\ 0 & 18 & 3 \end{bmatrix}$$

The commutative and associative laws of addition apply; i.e.,

$$\boldsymbol{A} + \boldsymbol{B} = \boldsymbol{B} + \boldsymbol{A} \tag{2.5}$$

$$(\boldsymbol{A} + \boldsymbol{B}) + \boldsymbol{C} = \boldsymbol{A} + (\boldsymbol{B} + \boldsymbol{C}) \tag{2.6}$$

Subtraction

Matrix subtraction is defined in a similar fashion. If two matrices $\boldsymbol{A}$ and $\boldsymbol{B}$ have the same number of rows and columns, the difference may be formed to obtain $\boldsymbol{D}$, where

$$\boldsymbol{A} - \boldsymbol{B} = \boldsymbol{D} \tag{2.7}$$

and the elements of $\boldsymbol{D}$ are defined by

$$d_{ij} = a_{ij} - b_{ij} \tag{2.8}$$

Equality

Matrices $\boldsymbol{A}$ and $\boldsymbol{B}$, both with dimensions $m \times n$, are said to be equal if their corresponding elements are equal; i.e.,

$$\boldsymbol{A} = \boldsymbol{B} \text{ if and only if } a_{ij} = b_{ij} \tag{2.9}$$

Multiplication

If k represents a scalar, i.e., a number, then

$$k\boldsymbol{A} = \boldsymbol{A}k = \{ka_{ij}\} \tag{2.10}$$

The above rule for scalar multiplication is simply that the scalar multiplies each element of a matrix.

The rule for multiplying two matrices together is somewhat more complex. If $\boldsymbol{A}$ has as many columns as $\boldsymbol{B}$ has rows, the product $\boldsymbol{AB}$ can be formed to yield $\boldsymbol{C}$; i.e.:

$$\boldsymbol{AB} = \boldsymbol{C} \tag{2.11}$$

where

$$c_{ij} = \sum_{k=1}^{n} a_{ik} b_{kj} \tag{2.12}$$

Note that from the above rule, the product matrix, $\boldsymbol{C}$, will have as many rows as $\boldsymbol{A}$ and as many columns as $\boldsymbol{B}$. The condition requiring as many columns in $\boldsymbol{A}$ as rows in $\boldsymbol{B}$ is called the *conformability condition.* $\boldsymbol{A}$ and $\boldsymbol{B}$ are said to be *conformable.*

Example 2.2 : Matrix multiplication.

Multiply the matrices

$$\boldsymbol{A} = \begin{bmatrix} 2 & 0 & -1 \\ 1 & 6 & 4 \end{bmatrix} \quad \boldsymbol{B} = \begin{bmatrix} 1 & -1 & 0 \\ -2 & 1 & 1 \\ 2 & -5 & 2 \end{bmatrix}$$

First, are they conformable? Yes, $\boldsymbol{A}$ has 3 columns and $\boldsymbol{B}$ 3 rows. According to Eq. (2.12) the resultant matrix will have 2 rows and 3 columns. Thus

$$\begin{aligned} c_{11} &= \sum_{k=1}^{3} a_{1k} b_{k1} = a_{11}b_{11} + a_{12}b_{21} + a_{13}b_{31} = 0 \\ c_{12} &= \sum_{k=1}^{3} a_{1k} b_{k2} = a_{11}b_{12} + a_{12}b_{22} + a_{13}b_{32} = 3 \end{aligned}$$

Continuing in this manner it can be shown that

$$\boldsymbol{AB} = \begin{bmatrix} 2 & 0 & -1 \\ 1 & 6 & 4 \end{bmatrix} \begin{bmatrix} 1 & -1 & 0 \\ -2 & 1 & 1 \\ 2 & -5 & 2 \end{bmatrix} = \begin{bmatrix} 0 & 3 & -2 \\ -3 & -15 & 14 \end{bmatrix}$$

The above is also referred to as *two-finger multiplication*, for reasons that will become clearer as the user's multiplication skills increase.

It should be noted that conformability of $\boldsymbol{A}$ and $\boldsymbol{B}$ in the order stated does not necessary imply conformability in the reverse order. The above example provides a good illustration; $\boldsymbol{A}$ is a 2×3 matrix and $\boldsymbol{B}$ is a 3×3 matrix. Thus $\boldsymbol{A}$ and $\boldsymbol{B}$ are conformable in that order. However, the order $\boldsymbol{BA}$ is not conformable and matrix multiplication is not defined. It should also be stated that even when two matrices $\boldsymbol{A}$ and $\boldsymbol{B}$ are conformable in either order, the resultant products are generally not equal; i.e.,

$$\boldsymbol{AB} \neq \boldsymbol{BA} \tag{2.13}$$

Example 2.3 : Commutativity and multiplication.

Consider

$$\boldsymbol{A} = \begin{bmatrix} 1 & 2 \\ -3 & 4 \end{bmatrix} \quad \boldsymbol{B} = \begin{bmatrix} 5 & -1 \\ -2 & 7 \end{bmatrix}$$

Then

$$\boldsymbol{AB} = \begin{bmatrix} 1 & 13 \\ -23 & 31 \end{bmatrix} \quad \boldsymbol{BA} = \begin{bmatrix} 8 & 6 \\ -23 & 24 \end{bmatrix}$$

which clearly indicates that $\boldsymbol{AB} \neq \boldsymbol{BA}$.

In general, matrix multiplication is not commutative. On the other hand, the lows of associativity and distributivity do hold provided the matrices involved are conformable. In other words,

$$(\boldsymbol{AB})\,\boldsymbol{C} = \boldsymbol{A}(\boldsymbol{BC}) \quad \text{associativity} \tag{2.14}$$

$$\boldsymbol{A}(\boldsymbol{B}+\boldsymbol{C}) = \boldsymbol{AB} + \boldsymbol{AC} \quad \text{distributivity} \tag{2.15}$$

$$(\boldsymbol{B}+\boldsymbol{C})\boldsymbol{A} = \boldsymbol{BA} + \boldsymbol{CA} \quad \text{distributivity} \tag{2.16}$$

In Eq. (2.15) the matrix sum $\boldsymbol{B} + \boldsymbol{C}$ is said to be *pre-multiplied* by $\boldsymbol{A}$. In Eq. (2.16) the sum $\boldsymbol{B} + \boldsymbol{C}$ is said to be *post-multiplied* by $\boldsymbol{A}$. Obviously, the other of multiplication is important.

2.1.3 Special (Square) Matrices

Null Matrix

The *null matrix* is a matrix which has all its elements equal to zero. It is usually denoted by $\mathbf{0}$. It is useful to note that two matrices $\boldsymbol{A}$ and $\boldsymbol{B}$ are said to be equal, if and only, if $\boldsymbol{A} - \boldsymbol{B} = \mathbf{0}$. One other point: The null matrix does not need to be square.

Diagonal Matrix

A *diagonal matrix* is one in which the only non-zero elements are located on the main diagonal. The main diagonal of a matrix consists of the elements a_{ii}. A diagonal matrix is indicated by

$$\boldsymbol{D} = \{d_{ij}\delta_{ij}\} = diag\,\lceil d_{11}, d_{22}, ..., d_{nn} \rfloor \tag{2.17}$$

where δ_{ij} is the *Kronecker delta* and is defined such that

$$\delta_{ij} = \begin{cases} 1 & if \quad i = j \\ 0 & otherwise \end{cases}$$

Example 2.4 : Diagonal matrix.

The matrix below is a diagonal matrix.

$$\boldsymbol{D} = \begin{bmatrix} 1 & 0 & 0 \\ 0 & 3 & 0 \\ 0 & 0 & -2 \end{bmatrix} = \begin{bmatrix} 1 & & zeroes \\ & 3 & \\ zeroes & & -2 \end{bmatrix} = diag\,\lceil 1, 3, -2 \rfloor$$

Post-multiplication by D

Post-multiplication of a matrix $\boldsymbol{A}$ by a diagonal matrix $\boldsymbol{D}$ is equivalent to an operation on the columns of $\boldsymbol{A}$. In other words, it multiplies all the elements in a given column by the same scalar.

Pre-multiplication by D

Pre-multiplication of a matrix $\boldsymbol{A}$ by a diagonal matrix $\boldsymbol{D}$ is equivalent to an operation on the rows of $\boldsymbol{A}$. In other words, all the elements in a given row are multiplied by the same scalar.

Example 2.5 : Post- and premultiplication.

$$\begin{bmatrix} a_{11} & a_{12} \\ a_{21} & a_{22} \end{bmatrix} \begin{bmatrix} d_{11} & 0 \\ 0 & d_{22} \end{bmatrix} = \begin{bmatrix} a_{11}d_{11} & a_{12}d_{22} \\ a_{21}d_{11} & a_{22}d_{22} \end{bmatrix} \quad Post\text{-}multiplication$$

$$\begin{bmatrix} d_{11} & 0 \\ 0 & d_{22} \end{bmatrix} \begin{bmatrix} a_{11} & a_{12} \\ a_{21} & a_{22} \end{bmatrix} = \begin{bmatrix} a_{11}d_{11} & a_{12}d_{11} \\ a_{21}d_{22} & a_{22}d_{22} \end{bmatrix} \quad Pre\text{-}multiplication$$

Identity or Unit Matrix

The *identity matrix* is a special type of diagonal matrix, in that each of the diagonal elements is equal to 1. It is usually denoted by $\boldsymbol{I}$.

Thus

$$\boldsymbol{I} = \{\ \delta_{ij}\ \} = \begin{bmatrix} 1 & 0 & 0 & .... & 0 \\ 0 & 1 & 0 & .... & 0 \\ 0 & 0 & 1 & .... & 0 \\ . & . & . & .... & . \\ 0 & 0 & 0 & .... & 1 \end{bmatrix} \tag{2.18}$$

The identity matrix has the very useful property of leaving one matrix upon which it operates unchanged; moreover, it commutes with any square matrix $\boldsymbol{A}$; i.e.,

$$\boldsymbol{IA} = \boldsymbol{AI} = \boldsymbol{A}$$

Triangular Matrix

A *triangular matrix* is a square matrix wherein the non-zero elements form a triangular array. If the elements below and to the left of the main diagonal are all zeros, it is called an *upper triangular matrix.* If the elements above and to the right of the main diagonal are all zeros, this is called a *lower triangular matrix.*

Example 2.6 : Triangular matrix.

$$\textit{Upper triangular matrix}: \begin{bmatrix} 2 & 5 & 7 \\ 0 & 1 & 3 \\ 0 & 0 & 4 \end{bmatrix}$$

$$\textit{Lower triangular matrix}: \begin{bmatrix} 3 & 0 & 0 \\ 8 & 2 & 0 \\ 1 & -3 & 4 \end{bmatrix}$$

Transposed Matrix

A *transposed matrix* is one where the rows and columns have been interchanged. It is denoted by a superscript T. Thus if $\boldsymbol{A} = \{a_{ij}\}$, then $\boldsymbol{A}^T = \{a_{ji}\}$. In Eq. (2.2), a column vector was indicated by $\boldsymbol{a}$. Within this context, $\boldsymbol{a}^T$ represents a row vector.

Example 2.7 : Transposed matrix.

$$\boldsymbol{A} = \begin{bmatrix} 1 & 5 & 18 \\ 7 & 2 & 3 \\ 19 & 25 & 31 \end{bmatrix}$$

$$\boldsymbol{A}^T = \begin{bmatrix} 1 & 7 & 19 \\ 5 & 2 & 25 \\ 18 & 3 & 31 \end{bmatrix}$$

Multiplication of Transposed Matrices

To find the transpose of a matrix which is a product of two or more matrices, one simply forms the product of the two or more transposed matrices in reverse order. That is, if $\boldsymbol{AB} = \boldsymbol{C}$ and $\boldsymbol{C}^T$ is required, then

$$\boldsymbol{C}^T = (\boldsymbol{AB})^T = \boldsymbol{B}^T\boldsymbol{A}^T \tag{2.19}$$

The proof is left to the reader as an exercise.

Symmetric Matrix

A square matrix is said to be *symmetric* if it is equal to its transpose, i.e.; if

$$\boldsymbol{A} = \boldsymbol{A}^T \text{ or } a_{ij} = a_{ji}$$

Example 2.8 : Symmetric matrix.

The following matrix is symmetric:

$$\boldsymbol{A} = \begin{bmatrix} 5 & 2 & 1 \\ 2 & -6 & 4 \\ 1 & 4 & 9 \end{bmatrix} = \boldsymbol{A}^T.$$

Skew–Symmetric Matrix

A *skew-symmetric matrix* is a matrix that is the negative of its transpose; i.e.,

$$\boldsymbol{A} = -\boldsymbol{A}^T \text{ or } a_{ij} = -a_{ji} \qquad (2.20)$$

Example 2.9 : Skew-symmetric.

The following matrix is skew–symmetric:

$$\boldsymbol{A} = \begin{bmatrix} 0 & 3 & 2 \\ -3 & 0 & -9 \\ -2 & 9 & 0 \end{bmatrix}$$

Orthogonal Matrix

An *orthogonal matrix* has the property that when multiplied by its transpose, the identity matrix is produced; i.e., if $\boldsymbol{A}$ is orthogonal,

$$\boldsymbol{A}\boldsymbol{A}^T = \boldsymbol{I}$$

Example 2.10 : Orthogonal matrix.

The following is an orthogonal matrix:

$$\boldsymbol{A} = \begin{bmatrix} \frac{1}{\sqrt{2}} & -\frac{1}{\sqrt{2}} \\ \frac{1}{\sqrt{2}} & \frac{1}{\sqrt{2}} \end{bmatrix}$$

Determinants

It is possible to associate with every square matrix $\boldsymbol{A}$ a scalar quantity called the *determinant* of $\boldsymbol{A}$. The determinant of the matrix $\boldsymbol{A}$ is written $det\boldsymbol{A}$ or $|\boldsymbol{A}|$. In relation to the matrix defined in Eq. (2.1), for $m = n$,

$$det\boldsymbol{A} = |\boldsymbol{A}| = \begin{vmatrix} a_{11} & a_{12} & \dots & a_{1n} \\ a_{21} & a_{22} & \dots & a_{2n} \\ \cdot & \cdot & \dots & \cdot \\ a_{n1} & a_{n2} & \dots & a_{nn} \end{vmatrix} \tag{2.21}$$

The determinant above is said to be of nth order, since it contains n rows and n columns. A square array of numbers enclosed by vertical lines as above denotes a determinant. It must not be confused with a matrix. Note that the value of $det\boldsymbol{A}$ is defined to be the sum of term such as

$$\sum (-1)^k a_{1a}a_{2b}a_{3c}\dots a_{nr} \tag{2.22}$$

Where the numbers a, b, c,...r are taken in some order and the summation is taken over all possible permutations of these numbers. There are $n!$ terms in the sum, since there are $n!$ possible permutations of the second subscript.

An important consequence of the definition is that each of the terms in the sum are the product of n elements of type a_{ij} in such a manner that exactly one element from a given row and a given column appears in each product. The sign of each term depends on the quantity k, which for a given term, is defined to be the number of inversions necessary to bring about a certain rearrangement for that term.

The rearrangements and inversions are obtained as follows. Suppose a term is of the form a_{1a}, a_{2b}, a_{3c}, ...,a_{nr}. Suppose that order of two of the adjacent a's are interchanged. That is called *inversion*. A finite number of such inversions may result in a rearrangement of the elements, so that the second, rather than the first subscripts are *normal*, i.e., arranged in sequence.

Example 2.11 : Evaluate determinant.

Evaluate the third-order determinant

$$det\boldsymbol{A} = \begin{vmatrix} a_{11} & a_{12} & a_{13} \\ a_{21} & a_{22} & a_{23} \\ a_{31} & a_{32} & a_{33} \end{vmatrix}$$

The $det\boldsymbol{A}$ is the sum of 3! terms. These are

(1) $a_{11}a_{22}a_{33}$ (4) $a_{12}a_{23}a_{31}$

(2) $a_{11}a_{23}a_{32}$ (5) $a_{13}a_{21}a_{32}$

(3) $a_{12}a_{21}a_{33}$ (6) $a_{13}a_{22}a_{31}$

To illustrate the above definition, use the product term $a_{13}a_{21}a_{32}$. Rearrangement of this term into "normal" order by successive inversions might take the form

$$a_{13}a_{21}a_{32} \quad \rightarrow \quad a_{21}a_{13}a_{32} \quad \rightarrow \quad a_{21}a_{32}a_{13}$$

so that $k = 2$; i.e., two inversions are needed. In a similar manner, values of k for the other terms are found to be

(1) 0 (4) 2

(2) 1 (5) 2

(3) 1 (6) 3

The number k is not unique; it depends on the efficiency of the rearranging process. The factor $(-1)^k$, however, does not depend on the rearrangement process. For determinants of fourth order and larger, this is a terribly cumbersome method of evaluation. Instead use is often made of certain basic properties to shorten the evaluation.

Basic Properties

Some basic properties of determinants, all depending on Eq. (2.22), are stated below.

1. If a row or column is totally composed of zeros, then the value of the determinant is zero.

2. If every element in a given row or column is multiplied by a scalar p, the value of the determinant is multiplied by p.

3. If two rows or two columns are interchanged, the sign of the determinant is changed.

4. If two rows or columns have a common ratio, the value of the determinant is zero.

5. If p times a given row or column is added (or subtracted) from another row or column, the value of the determinant is unchanged.

6. The determinant of the product of two matrices is equal to the product of the determinant of the two matrices; i.e., if $\boldsymbol{AB} = \boldsymbol{C}$, then $|\boldsymbol{A}||\boldsymbol{B}| = |\boldsymbol{C}|$.

Example 2.12 : Numerical calculations.

Evaluate the fourth-order determinant.

$$|\boldsymbol{A}| = \begin{vmatrix} 2 & 1 & 4 & 3 \\ 6 & -1 & 2 & -4 \\ 3 & -2 & 5 & 1 \\ -5 & 6 & 4 & -1 \end{vmatrix}$$

Using property 5, subtract 6/2, 3/2 and -5/2 times the first row from the second, third and fourth rows, respectively. Thus

$$|\boldsymbol{A}| = \begin{vmatrix} 2 & 1 & 4 & 3 \\ 0 & -4 & -10 & -13 \\ 0 & -3.5 & -1 & -3.5 \\ 0 & 8.5 & 14 & 6.5 \end{vmatrix}$$

Next, subtract 3.5/4 and -8.5/4 times the second row from the third and fourth row, respectively.

The result is

$$|\boldsymbol{A}| = \begin{vmatrix} 2 & 1 & 4 & 3 \\ 0 & -4 & -10 & -13 \\ 0 & 0 & 7.75 & 7.875 \\ 0 & 0 & -7.25 & -21.125 \end{vmatrix}$$

Finally subtract - 7.25/7.75 times the third row from the fourth row to produce.

$$|\boldsymbol{A}| = \begin{vmatrix} 2 & 1 & 4 & 3 \\ 0 & -4 & -10 & -13 \\ 0 & 0 & 7.75 & 7.875 \\ 0 & 0 & 0 & -853/62 \end{vmatrix}$$

Consequently,

$$|\boldsymbol{A}| = 853$$

The determinant above is said to be a *triangular determinant*: Elements below the main diagonal are all zero. The property of such determinants is as follows:

7. The determinant of a triangular matrix is equal to the product of the diagonal elements.

Minors and Cofactors

If p rows and p columns are deleted from an nth order determinant, the resulting determinant of $n-p$ rows and $n-p$ columns is called a pth *minor.* In particular, if p is 1, the determinant of order $n-1$ is called a first order minor or simply a minor. The notation $|\boldsymbol{M}_{ij}|$ is used to designate the minor resulting from the deletion of the ith row and the jth column from $|\boldsymbol{A}|$ and is said to belong to the element a_{ij}. Minors whose diagonal elements are also diagonal element of $|\boldsymbol{A}|$ are called *principal minors* of $|\boldsymbol{A}|$. $|\boldsymbol{M}_{ii}|$ is such a principal minor. The scalar quantity denoted by C_{ij} and defined by the equation.

$$C_{ij} = (-1)^{i+j} |\boldsymbol{M}_{ij}| \tag{2.23}$$

is called the *cofactor* of the element a_{ij}. Note that this is simply a minor with a sign attached.

Adjoint Matrix

The *adjoint matrix* is the transpose of the matrix formed by replacing each element by its cofactor. It is denoted by *adj* $\boldsymbol{A}$. If $\boldsymbol{A}$ is a square matrix and C_{ij} the cofactor of a_{ij}, then the adjoint matrix is

$$adj\,\boldsymbol{A} = \{C_{ji}\} \tag{2.24}$$

Example 2.13 : Ajoint matrix.

Let us find the adjoint matrix for

$$\boldsymbol{A} = \begin{bmatrix} 2 & 1 & -1 \\ 1 & 2 & 3 \\ 3 & 0 & 4 \end{bmatrix}$$

The matrix of cofactors of $\boldsymbol{A}$ is

$$\{C_{ij}\} = \begin{bmatrix} 8 & 5 & -6 \\ -4 & 11 & 3 \\ 5 & -7 & 3 \end{bmatrix}$$

Thus, the adjoint of $\boldsymbol{A}$ is

$$adj\,\boldsymbol{A} = \begin{bmatrix} 8 & -4 & 5 \\ 5 & 11 & -7 \\ -6 & 3 & 3 \end{bmatrix}$$

Non-Singular Matrix

If $\boldsymbol{A}$ is a matrix for which $det\boldsymbol{A} \neq 0$, it is called *non-singular*. If $det\boldsymbol{A} = 0$, it is called a singular matrix.

Inverse Matrix

The *inverse matrix*, denoted by a superscript (-1) is defined by the property

$$\boldsymbol{A}\boldsymbol{A}^{-1} = \boldsymbol{A}^{-1}\boldsymbol{A} = \boldsymbol{I} \tag{2.25}$$

In words, an inverse of a matrix is one that when multiplied by the matrix produces the identify matrix.

Existence of the Inverse

Every square matrix $\boldsymbol{A}$ possesses an inverse provided $\boldsymbol{A}$ is non-singular. Conversely, if $\boldsymbol{A}$ is singular, the inverse does not exist. To show this, note that since $\boldsymbol{A}\boldsymbol{R} = \boldsymbol{I}$, the product of the determinants of $\boldsymbol{A}$ and $\boldsymbol{R}$ must equal 1; i.e.,

$$|\boldsymbol{A}||\boldsymbol{R}| = 1$$

This cannot be if $|\boldsymbol{A}| = 0$

Construction of the Inverse

It can be proved that the inverse may be calculated from:

$$\boldsymbol{A}^{-1} = \frac{adj\,\boldsymbol{A}}{|\boldsymbol{A}|} \tag{2.26}$$

Example 2.14 : Find inverse matrix.

Find the inverse of $\boldsymbol{A}$

$$\boldsymbol{A} = \begin{bmatrix} 2 & 1 & -1 \\ 1 & 2 & 3 \\ 3 & 0 & 4 \end{bmatrix}$$

Note that the adjoint of $\boldsymbol{A}$ has previously been obtained in Example 2.13. Also,

$$|\boldsymbol{A}| = 27$$

hence, using Eq. (2.26), gives the results

$$\boldsymbol{A}^{-1} = \frac{adj\,\boldsymbol{A}}{|\boldsymbol{A}|} = \frac{1}{27}\begin{bmatrix} 8 & -4 & 5 \\ 5 & 11 & -7 \\ -6 & 3 & 3 \end{bmatrix}$$

The reader may verify that the above is indeed a proper inverse by computing the product of $\boldsymbol{A}$ and $\boldsymbol{A}^{-1}$.

Products of Inverse Matrices

The inverse of a product of matrices is equal to the product of the inverse in reverse order. That is, if $\boldsymbol{AB} = \boldsymbol{C}$, then $\boldsymbol{C}^{-1} = \boldsymbol{B}^{-1}\boldsymbol{A}^{-1}$. The proof is left to the reader as an exercise. Note that the above conclusion is valid for any number of products.

Partitioned Matrices

The matrix elements up to now have been considered as single numbers, they may however be matrices themselves. Another way to say this is that a matrix whose elements are numbers may be partitioned so that the matrix has a smaller number of "elements", these elements being matrices. For example, by drawing the lines in the 3×3 matrix $\boldsymbol{A}$ below, a 2×2 matrix may be formed whose elements are themselves matrices.

$$\boldsymbol{A} = \left[\begin{array}{cc|c} a_{11} & a_{12} & a_{13} \\ a_{21} & a_{22} & a_{23} \\ \hline a_{31} & a_{32} & a_{33} \end{array}\right] = \begin{bmatrix} \boldsymbol{A}_1 & \boldsymbol{\alpha}_2 \\ \boldsymbol{\alpha}_3^T & a_{33} \end{bmatrix} \tag{2.27}$$

The partitioned matrices obey the properties of matrices in general, provided it is always kept in mind that the elements are themselves matrices. Suppose a $\boldsymbol{B}$ matrix is partitioned in the following manner:

$$\boldsymbol{B} = \left[\begin{array}{c|cc} b_{11} & b_{12} & b_{13} \\ b_{21} & b_{22} & b_{23} \\ \hline b_{31} & b_{32} & b_{33} \end{array}\right] = \begin{bmatrix} \boldsymbol{\beta}_1 & \boldsymbol{B}_1 \\ b_{31} & \boldsymbol{\beta}_2^T \end{bmatrix} \tag{2.28}$$

The product of $\boldsymbol{A}$ and $\boldsymbol{B}$ can then be expressed as:

$$\boldsymbol{AB} = \begin{bmatrix} \boldsymbol{A}_1\boldsymbol{\beta}_1 + \boldsymbol{\alpha}_2 b_{31} & \boldsymbol{A}_1\boldsymbol{B}_1 + \boldsymbol{\alpha}_2\boldsymbol{\beta}_2^T \\ \boldsymbol{\alpha}_3^T\boldsymbol{\beta}_1 + a_{33}b_{31} & \boldsymbol{\alpha}_3^T\boldsymbol{B}_1 + a_{33}\boldsymbol{\beta}_2^T \end{bmatrix} \tag{2.29}$$

In general, the product of two partitioned conformable matrices can be expressed in terms of the sub-matrices only if the proposed products involve only conformable terms. The reader may verify that the terms of Eq. (2.29) are conformable.

2.1.4 Vector Spaces

A vector with n elements may be represented by a point in an n-dimensional space. This is done by using a set of n coordinate axes. When $n = 2$ or 3, the geometric interpretations are clear. When $n > 3$, the geometric interpretation is not clear at all; however, the association with a finite-dimensional space is retained.

In the above, the n coordinate axes may be taken to be almost any set of n vectors. More commonly, however, they are taken to be orthogonal to each other and of unit length. In three-dimensional space, this corresponds to the usual x, y and z-axes, where a set of unit vectors could be designated by

$$\begin{aligned} \boldsymbol{e}_1^T &= (1,0,0) \\ \boldsymbol{e}_2^T &= (0,1,0) \\ \boldsymbol{e}_3^T &= (0,0,1) \end{aligned} \tag{2.30}$$

It can be seen that any vector in three dimensions may be represented as a linear combination of the unit vectors. If a vector is defined as

$$\boldsymbol{x}^T = (5,\ 3,\ -2)$$

it can be written in terms of the unit vectors as

$$\boldsymbol{x}^T = 5\boldsymbol{e}_1^T + 3\boldsymbol{e}_2^T - 2\boldsymbol{e}_3^T$$

Scalar Product

The *scalar* or *inner product* of two real vectors $\boldsymbol{x}$ and $\boldsymbol{y}$, designated by either $\boldsymbol{x}^T\boldsymbol{y}$ or $\boldsymbol{x} \cdot \boldsymbol{y}$, is defined as

$$\boldsymbol{x}^T\boldsymbol{y} = \boldsymbol{x} \cdot \boldsymbol{y} = x_1y_1 + x_2y_2 + \ldots + x_ny_n \tag{2.31}$$

where $\boldsymbol{x}^T$ is $1 \times n$ and $\boldsymbol{y}$ is $n \times 1$.

Example 2.15 : Scalar product.

Determine the inner product for

$$\begin{aligned} \boldsymbol{x}^T &= (5,3,-2) \quad \boldsymbol{y}^T = (-4,-7,8) \\ \textit{Thus} \quad \boldsymbol{x}^T\boldsymbol{y} &= [5 \quad 3 \ -2] \begin{bmatrix} -4 \\ -7 \\ 8 \end{bmatrix} = -20 - 21 - 16 = -57 \end{aligned}$$

Observe that the product is a *scalar* quantity, as the name of the product implies.

Orthogonal Vectors

Two vectors $\boldsymbol{x}$ and $\boldsymbol{y}$ are said to be *orthogonal* if their scalar product is zero; i.e.,

$$\boldsymbol{x}^T\boldsymbol{y} = \boldsymbol{x}\cdot\boldsymbol{y} = 0 \tag{2.32}$$

Example 2.16 : Orthogonal vectors.

$$\begin{aligned}\text{If}\quad \boldsymbol{x}^T &= (1,0,0)\ \text{ and }\ \boldsymbol{y}^T = (0,1,0)\,,\\ \boldsymbol{x}^T\boldsymbol{y} &= \begin{bmatrix}1 & 0 & 0\end{bmatrix}\begin{bmatrix}0\\1\\0\end{bmatrix} = 0\end{aligned}$$

Thus $\boldsymbol{x}$ and $\boldsymbol{y}$ are orthogonal.

Length of Vector

The length of a real vector, denoted $||\boldsymbol{x}||$, is defined as

$$||\boldsymbol{x}|| = \sqrt{\boldsymbol{x}^T\boldsymbol{x}} = \sqrt{x_1^2 + x_2^2 + ... + x_n^2} \tag{2.33}$$

Unit Vectors

A *unit vector* has unit length. If $\boldsymbol{e}$ is a unit vector, $||\boldsymbol{e}|| = 1$. A unit vector can be obtained from any vector $\boldsymbol{x}$ by simply dividing by $||\boldsymbol{x}||$ i.e.,

$$\hat{\boldsymbol{x}} = \frac{\boldsymbol{x}}{||\boldsymbol{x}||}$$

Example 2.17 : Unit vectors.

If $\boldsymbol{x}^T = \left(2, -4, \sqrt{5}\right)$ determine the corresponding unit vector. The length of $\boldsymbol{x}$ or the *norm* of $\boldsymbol{x}$ is given by

$$||\boldsymbol{x}|| = \sqrt{\boldsymbol{x}\cdot\boldsymbol{x}} = \sqrt{4+16+5} = 5$$

Hence

$$\hat{\boldsymbol{x}} = \frac{1}{5}\begin{bmatrix}2\\-4\\\sqrt{5}\end{bmatrix}$$

Orthonormal Vectors

Two vectors $\boldsymbol{x}$ and $\boldsymbol{y}$ are said to be *orthonormal* if they are orthogonal and of unit length.

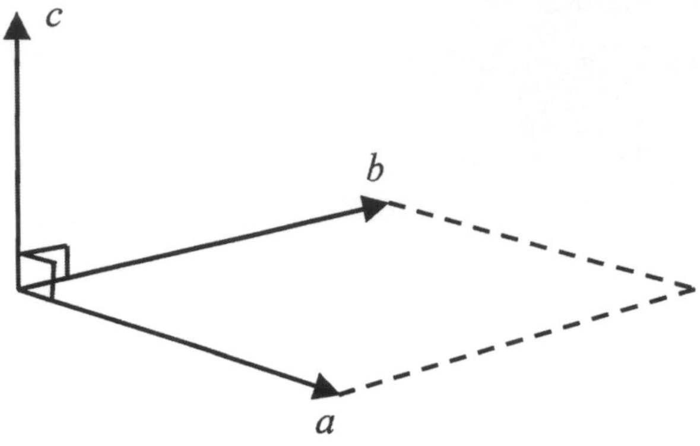

Figure 2.1: Vector product.

Vector Product

Various applications suggest the introduction of another kind of vector multiplication in which the product of two vectors is again a vector. This so-called *vector product* or *cross product* of two vectors $\boldsymbol{a}$ and $\boldsymbol{b}$ both with geometric interpretation can be formed to yield $\boldsymbol{c}$; i.e.,

$$\boldsymbol{a} \times \boldsymbol{b} = \boldsymbol{c} \tag{2.34}$$

Where $\boldsymbol{c}$ is a vector which is defined as follows.

If $\boldsymbol{a}$ and $\boldsymbol{b}$ have the same or opposite direction or one of these vectors is the zero vector, then $\boldsymbol{c} = \boldsymbol{0}$. In any other case, $\boldsymbol{c}$ is the vector whose length is equal to the area of the parallelogram with $\boldsymbol{a}$ and $\boldsymbol{b}$ as adjacent sides and whose direction is perpendicular to both $\boldsymbol{a}$ and $\boldsymbol{b}$ and is such that $\boldsymbol{a}$, $\boldsymbol{b}$ and $\boldsymbol{c}$, in this order forms a right-handed triple or right-handed triad, as shown in Fig. 2.1.

The term *right-handed* comes from the fact that when forming the three vectors $\boldsymbol{a}$, $\boldsymbol{b}$ and $\boldsymbol{c}$ by the thumb, index finger and middle finger, respectively, of the right hand, the direction of vector $\boldsymbol{c}$ will be outwards along the middle finger. The parallelogram with $\boldsymbol{a}$ and $\boldsymbol{b}$ as adjacent sides has the area $\|\boldsymbol{a}\|\ \|\boldsymbol{b}\|\ \sin\gamma$ where γ is the angle between vector $\boldsymbol{a}$ and $\boldsymbol{b}$. We thus obtain

$$\|\boldsymbol{c}\| = \|\boldsymbol{a}\|\ \|\boldsymbol{b}\| \sin\gamma \tag{2.35}$$

Let $\boldsymbol{a} \times \boldsymbol{b} = \boldsymbol{c}$ and let $\boldsymbol{b} \times \boldsymbol{a} = \boldsymbol{d}$. Then, by definition $\|\boldsymbol{c}\| = \|\boldsymbol{d}\|$, and in order that $\boldsymbol{b}$, $\boldsymbol{a}$ and $\boldsymbol{d}$ form a right-handed triple we must have $\boldsymbol{d} = -\boldsymbol{c}$, see Fig. 2.1. Hence,

$$\boldsymbol{b} \times \boldsymbol{a} = -\left(\boldsymbol{a} \times \boldsymbol{b}\right); \tag{2.36}$$

that is, cross-multiplication of vectors is not commutative but *anticommutative.* The order of the factors in a vector product is, therefore, of great importance and must be carefully observed. From the definition it follows that for any constant k,

$$(k\boldsymbol{a}) \times \boldsymbol{b} = k(\boldsymbol{a} \times \boldsymbol{b}) = \boldsymbol{a} \times (k\boldsymbol{b}) \tag{2.37}$$

Furthermore cross-multiplication is distributive with respect to vector addition, that is,

$$\boldsymbol{a} \times (\boldsymbol{b} + \boldsymbol{c}) = (\boldsymbol{a} \times \boldsymbol{b}) + (\boldsymbol{a} \times \boldsymbol{c}) \tag{2.38}$$

$$(\boldsymbol{a} + \boldsymbol{b}) \times \boldsymbol{c} = (\boldsymbol{a} \times \boldsymbol{c}) + (\boldsymbol{b} \times \boldsymbol{c}) \tag{2.39}$$

The proof is left to the reader. More complicated cross products may be considered, and we want to mention that

$$\boldsymbol{a} \times (\boldsymbol{b} \times \boldsymbol{c}) \neq (\boldsymbol{a} \times \boldsymbol{b}) \times \boldsymbol{c},$$

ordinarily.

The scalar product (Eq. 2.31) may also be written

$$\boldsymbol{a} \cdot \boldsymbol{b} = \|\boldsymbol{a}\| \|\boldsymbol{b}\| \cos \gamma$$

Combining this with Eq. (2.35) it follows that

$$\|\boldsymbol{c}\|^2 = \|\boldsymbol{a}\|^2 \|\boldsymbol{b}\|^2 \sin^2\gamma = \|\boldsymbol{a}\|^2 \|\boldsymbol{b}\|^2 \left(1 - \cos^2\gamma\right) = (\boldsymbol{a} \cdot \boldsymbol{a})(\boldsymbol{b} \cdot \boldsymbol{b}) - (\boldsymbol{a} \cdot \boldsymbol{b})^2$$

Hence we obtain the formula

$$\|\boldsymbol{a} \times \boldsymbol{b}\| = \sqrt{(\boldsymbol{a} \cdot \boldsymbol{a})(\boldsymbol{b} \cdot \boldsymbol{b}) - (\boldsymbol{a} \cdot \boldsymbol{b})^2} \tag{2.40}$$

which is convenient for computing the length of a vector product.

Vector Product in Terms of Components

The geometric vectors $\boldsymbol{a}$ and $\boldsymbol{b}$ may be written

$$\boldsymbol{a} = \begin{bmatrix} a_1 \\ a_2 \\ a_3 \end{bmatrix} \qquad \boldsymbol{b} = \begin{bmatrix} b_1 \\ b_2 \\ b_3 \end{bmatrix}$$

Referring to the orthonormal vectors of Eq. (2.30) the vector product may be written

$$\boldsymbol{c} \;=\; \boldsymbol{a} \times \boldsymbol{b} = \begin{vmatrix} a_2 & a_3 \\ b_2 & b_3 \end{vmatrix} \boldsymbol{e}_1 + \begin{vmatrix} a_3 & a_1 \\ b_3 & b_1 \end{vmatrix} \boldsymbol{e}_2 + \begin{vmatrix} a_1 & a_2 \\ b_1 & b_2 \end{vmatrix} \boldsymbol{e}_3$$

$$= c_1\boldsymbol{e}_1 + c_2\boldsymbol{e}_2 + c_3\boldsymbol{e}_3 \tag{2.41}$$

where

$$\begin{aligned} c_1 &= a_2b_3 - b_2a_3 \\ c_2 &= a_3b_1 - b_3a_1 \\ c_3 &= a_1b_2 - b_1a_2 \end{aligned} \tag{2.42}$$

and

$$\boldsymbol{c} = \begin{bmatrix} c_1 \\ c_2 \\ c_3 \end{bmatrix}$$

2.2 Transformations and Rotations

2.2.1 Coordinate Transformation

The principle of coordinate transformation will first be demonstrated by rotating an $x - y$ coordinate system in two dimensions, see Fig. 2.2. By geometric inspection the following equations may be set up.

$$\begin{aligned} v_x &= v_{x'}\cos\gamma - v_{y'}sin\gamma \\ v_y &= v_{x'}sin\gamma + v_{y'}\cos\gamma \end{aligned} \tag{2.43}$$

In matrix form this may be written

$$\begin{bmatrix} v_x \\ v_y \end{bmatrix} = \begin{bmatrix} \cos\gamma & -\sin\gamma \\ \sin\gamma & \cos\gamma \end{bmatrix} \begin{bmatrix} v_{x'} \\ v_{y'} \end{bmatrix} \tag{2.44}$$

and in matrix and vector symbols:

$$\boldsymbol{v} = \boldsymbol{A}\boldsymbol{v}' \tag{2.45}$$

Example 2.18 : Rotate coordinate system.

Rotate the coordinate system 90° and calculate the coordinates in the original coordinate system for a point attached to the rotated x'-axis 5 units in the positive direction on the axis. Referring to Eqs. (2.44) and (2.45) the vector $\boldsymbol{v}$ is

$$\boldsymbol{v} = \boldsymbol{A} \cdot \boldsymbol{v}' = \begin{bmatrix} 0 & -1 \\ 1 & 0 \end{bmatrix} \begin{bmatrix} 5 \\ 0 \end{bmatrix} = \begin{bmatrix} 0 \\ 5 \end{bmatrix}$$

That is, the transformation rotates the point to the positive y-axis of the original coordinate system.

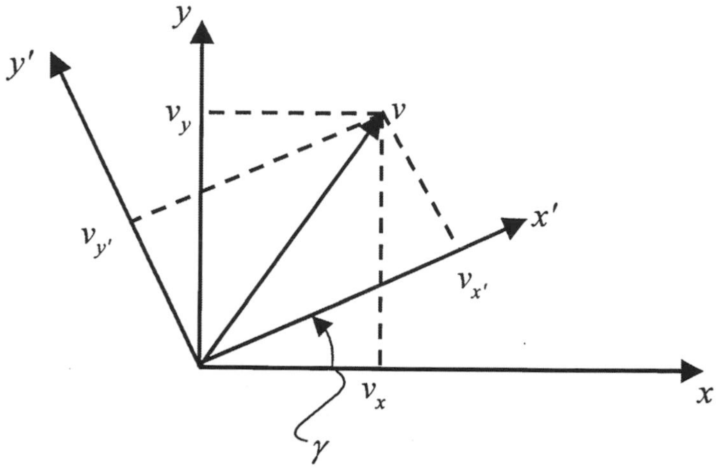

Figure 2.2: Coordinate Transformation.

The 4 × 4 Transformation Matrix (Homogeneous Coordinates)

Eqs. (2.44) and (2.45) only handle rotation of coordinate systems in two dimensions. In general a tool is needed for handling rotations and translations in 3D space, and this tool is the 4 × 4 transformation matrix $\boldsymbol{T}$:

$$\boldsymbol{T} = \begin{bmatrix} t_{11} & t_{12} & t_{13} & t_{14} \\ t_{21} & t_{22} & t_{23} & t_{24} \\ t_{31} & t_{32} & t_{33} & t_{34} \\ t_{41} & t_{42} & t_{43} & t_{44} \end{bmatrix} = \begin{bmatrix} \boldsymbol{R} & \boldsymbol{S} \\ \boldsymbol{P}^T & U \end{bmatrix} \tag{2.46}$$

where $\boldsymbol{R}$ is the (3 × 3) direction cosines, $\boldsymbol{S}$ is the (3 × 1) translation vector, $\boldsymbol{P}^T$ is the (3 × 1) perspective vector, and U is the uniform scaling factor.

Direction Cosines

The 3 × 3 *direction cosine* matrix $\boldsymbol{R}$ is simply the extension of matrix $\boldsymbol{A}$ in Eq. (2.45) to handle rotations in 3D space. Rotating the 3D coordinate system at an angle γ in a positive direction around the z-axis is written in matrix form, see Eq. (2.44).

$$\boldsymbol{R}_z = \begin{bmatrix} \cos\gamma & -\sin\gamma & 0 \\ \sin\gamma & \cos\gamma & 0 \\ 0 & 0 & 1 \end{bmatrix} \tag{2.47}$$

Positive rotation around an axis is defined as clockwise looking outward along the axis. Rotating an angle α and β around the x- and y-axis, respectively, is written in matrix form as follows:

$$\boldsymbol{R}_x = \begin{bmatrix} 1 & 0 & 0 \\ 0 & \cos\alpha & -\sin\alpha \\ 0 & \sin\alpha & \cos\alpha \end{bmatrix} \tag{2.48}$$

$$R_y = \begin{bmatrix} \cos\beta & 0 & \sin\beta \\ 0 & 1 & 0 \\ -\sin\beta & 0 & \cos\beta \end{bmatrix} \tag{2.49}$$

To generate the transformation matrix for rotation around two or three axes, matrix multiplication (not addition) of the matrices above is needed, and as matrix multiplication is not commutative the order of multiplication for the matrices above are crucial for the resulting transformation. The direction cosine matrix may also be written

$$\boldsymbol{R} = \begin{bmatrix} \hat{\boldsymbol{v}}_x \; \hat{\boldsymbol{v}}_y \; \hat{\boldsymbol{v}}_z \end{bmatrix} \tag{2.50}$$

where $\hat{\boldsymbol{v}}_x$, $\hat{\boldsymbol{v}}_y$ and $\hat{\boldsymbol{v}}_z$ are orthogonal unit vectors (orthonormal) along the x-, y- and z-axis, respectively, of the transformed coordinate system referred to the reference coordinate system.

Example 2.19 : Consecutive rotations.

Rotate the coordinate system 90° about the y-axis to the $\boldsymbol{v}_1$ position and further 90° about the z'-axis to the $\boldsymbol{v}_2$ position. Calculate the coordinates for a point attached to the x'-axis 5 units in positive direction on the axis, and that follows the rotations.

First rotate about the y-axis, $\boldsymbol{R}_1 = \boldsymbol{R}_y$, refer to Eq. (2.49):

$$\boldsymbol{v}_1 = \boldsymbol{R}_1\boldsymbol{v}' = \begin{bmatrix} 0 & 0 & 1 \\ 0 & 1 & 0 \\ -1 & 0 & 0 \end{bmatrix} \begin{bmatrix} 5 \\ 0 \\ 0 \end{bmatrix} = \begin{bmatrix} 0 \\ 0 \\ -5 \end{bmatrix}$$

That is, the rotation about the y-axis rotates the point to the negative z-axis of the original coordinate system.

The transformation matrix for the combined rotations may be calculated from

$$\boldsymbol{R}_2 = \boldsymbol{R}_y \cdot \boldsymbol{R}_z = \begin{bmatrix} 0 & 0 & 1 \\ 0 & 1 & 0 \\ -1 & 0 & 0 \end{bmatrix} \begin{bmatrix} 0 & -1 & 0 \\ 1 & 0 & 0 \\ 0 & 0 & 1 \end{bmatrix} = \begin{bmatrix} 0 & 0 & 1 \\ 1 & 0 & 0 \\ 0 & 1 & 0 \end{bmatrix}$$

The combined rotation of the point on the x'-axis may then be calculated from

$$\boldsymbol{v}_2 = \boldsymbol{R}_2\boldsymbol{v}' = \begin{bmatrix} 0 & 0 & 1 \\ 1 & 0 & 0 \\ 0 & 1 & 0 \end{bmatrix} \begin{bmatrix} 5 \\ 0 \\ 0 \end{bmatrix} = \begin{bmatrix} 0 \\ 5 \\ 0 \end{bmatrix}$$

That is, the combined rotation moved the point to the positive y-axis of the original coordinate system. These transformations are of course valid for any angles of rotation, 90° rotations are chosen only to simplify the calculations in the example. The reader is strongly recommended to draw the different coordinate systems and the transformed point in the different positions to gain confidence in coordinate transformation.

Translation Transformation

The *translation transformation* vector $\boldsymbol{S}$ from Eq. (2.46) is the translational part of the general (4×4) transformation matrix.

$$\boldsymbol{S}^T = [X\; Y\; Z]$$

The direction cosine part of the matrix handles the rotations in 3D space between coordinate systems, while the translational part $\boldsymbol{S}$ takes care of the translations of the origin of the coordinate systems. Together, these two parts of the matrix uniquely define the general transformation between coordinate systems in space.

Perspective and Scaling

The *perspective transformation* part $\boldsymbol{P}$ of the matrix (Eq. 2.46) is used, as the name indicates, to generate perspective views of a transformed body. For our purpose, this part of the matrix is not used, and is therefore set equal to the zero vector: $\boldsymbol{P}^T = \boldsymbol{0}$.

The uniform scaling factor for a transformed body, U, is for our purpose set equal to unity, that is $U = 1$. For more details, refer to textbooks on computer graphics.

Modified 4×4 Transformation

The modified transformation matrix is shown below:

$$\boldsymbol{T} = \begin{bmatrix} \boldsymbol{R} & \boldsymbol{S} \\ \boldsymbol{0} & 1 \end{bmatrix} = \begin{bmatrix} R_{11} & R_{12} & R_{13} & X \\ R_{21} & R_{22} & R_{23} & Y \\ R_{31} & R_{32} & R_{33} & Z \\ 0 & 0 & 0 & 1 \end{bmatrix} \tag{2.51}$$

The general transformation of coordinates and coordinate systems is then

$$\boldsymbol{v}_h = \boldsymbol{T}\boldsymbol{v}'_h \tag{2.52}$$

where $\boldsymbol{v}_h$ and $\boldsymbol{v}'_h$ are (4×1) vectors of Homogeneous Coordinates:

$$\boldsymbol{v}_h = \begin{bmatrix} v_x \\ v_y \\ v_z \\ 1 \end{bmatrix} \tag{2.53}$$

The position vector is now expanded from 3 to 4 where the fourth component is the counterpart to the scaling factor U of the transformation matrix, Eq. (2.46). For our purpose the fourth row of the transformation matrix (Eq. 2.51) is fixed to (0, 0, 0, 1) and the fourth element of the position vector is fixed to 1.

Example 2.20 : Rotate and translate.

Rotate the coordinate system 90° about the y-axis to the $\boldsymbol{v}_{h1}$ position and further translate 4 units in positive z' direction to the $\boldsymbol{v}_{h2}$ position. Calculate the coordinates for a point attached to the x'-axis 5 units in the positive direction on the axis, and that follows the rotation. The first transformation is exactly the same as the first transformation of Example 2.19:

$$\boldsymbol{v}_{h1} = \boldsymbol{T}_1\boldsymbol{v}'_h = \begin{bmatrix} 0 & 0 & 1 & 0 \\ 0 & 1 & 0 & 0 \\ -1 & 0 & 0 & 0 \\ 0 & 0 & 0 & 1 \end{bmatrix} \begin{bmatrix} 5 \\ 0 \\ 0 \\ 1 \end{bmatrix} = \begin{bmatrix} 0 \\ 0 \\ -5 \\ 1 \end{bmatrix}$$

As previously, the point is moved to the negative z-axis of the original coordinate system. The transformation matrix for the combined rotation and translation is thus

$$\boldsymbol{T}_2 = \begin{bmatrix} 0 & 0 & 1 & 0 \\ 0 & 1 & 0 & 0 \\ -1 & 0 & 0 & 0 \\ 0 & 0 & 0 & 1 \end{bmatrix} \begin{bmatrix} 1 & 0 & 0 & 0 \\ 0 & 1 & 0 & 0 \\ 0 & 0 & 1 & 4 \\ 0 & 0 & 0 & 1 \end{bmatrix} = \begin{bmatrix} 0 & 0 & 1 & 4 \\ 0 & 1 & 0 & 0 \\ -1 & 0 & 0 & 0 \\ 0 & 0 & 0 & 1 \end{bmatrix}$$

and the transformed point is calculated from:

$$\boldsymbol{v}_{h2} = \boldsymbol{T}_2 \cdot \boldsymbol{v}'_h = \begin{bmatrix} 0 & 0 & 1 & 4 \\ 0 & 1 & 0 & 0 \\ -1 & 0 & 0 & 0 \\ 0 & 0 & 0 & 1 \end{bmatrix} \begin{bmatrix} 5 \\ 0 \\ 0 \\ 1 \end{bmatrix} = \begin{bmatrix} 4 \\ 0 \\ -5 \\ 1 \end{bmatrix}$$

That is, the rotation about y followed by translation along z' moves the point on the x'-axis to position (4, 0, -5) referred to the original coordinate system. Drawing the different coordinate systems and the transformed point in the different positions is also recommended for this example.

Compact Matrix Format

As shown above, the fourth row of the 4×4 transformation matrix is constant and equal to (0, 0, 0, 1) and the fourth element of the position vector is constant and equal to (1). Consequently, the fourth row of the matrix and the fourth element of the vector does not have to be stored. That is, the transformation matrix may be stored, refer to Eq. (2.51):

$$\boldsymbol{t} = [\boldsymbol{R}\ \boldsymbol{S}] = \begin{bmatrix} R_{11} & R_{12} & R_{13} & X \\ R_{21} & R_{22} & R_{23} & Y \\ R_{31} & R_{32} & R_{33} & Z \end{bmatrix} \tag{2.54}$$

while the position vector is stored, refer to Eq. (2.53):

$$\boldsymbol{v} = \begin{bmatrix} v_x \\ v_y \\ v_z \end{bmatrix} \tag{2.55}$$

Reduced Matrix - Matrix Multiplication

Full matrix multiplication gives:

$$\boldsymbol{T} = \boldsymbol{T}_1\boldsymbol{T}_2 = \begin{bmatrix} \boldsymbol{R}_1 & \boldsymbol{S}_1 \\ \boldsymbol{0} & 1 \end{bmatrix}\begin{bmatrix} \boldsymbol{R}_2 & \boldsymbol{S}_2 \\ \boldsymbol{0} & 1 \end{bmatrix} = \begin{bmatrix} \boldsymbol{R}_1\boldsymbol{R}_2 & \boldsymbol{R}_1\boldsymbol{S}_2 + \boldsymbol{S}_1 \\ \boldsymbol{0} & 1 \end{bmatrix} \tag{2.56}$$

That is, modified multiplication gives, refer to Eq. (2.54):

$$\boldsymbol{R} = \boldsymbol{R}_1\boldsymbol{R}_2 \quad \boldsymbol{S} = \boldsymbol{R}_1\boldsymbol{S}_2 + \boldsymbol{S}_1 \tag{2.57}$$

Reduced Matrix - Vector Multiplication

Full matrix - vector multiplication gives:

$$\boldsymbol{v}_h = \boldsymbol{T}\boldsymbol{v}'_h = \begin{bmatrix} \boldsymbol{R} & \boldsymbol{S} \\ \boldsymbol{0} & 1 \end{bmatrix}\begin{bmatrix} \boldsymbol{v}' \\ 1 \end{bmatrix} = \begin{bmatrix} \boldsymbol{R}\boldsymbol{v}' + \boldsymbol{S} \\ 1 \end{bmatrix} \tag{2.58}$$

That is, modified matrix-vector multiplication gives, refer to Eq. (2.58):

$$\boldsymbol{v} = \boldsymbol{R}\boldsymbol{v}' + \boldsymbol{S} \tag{2.59}$$

Reduced Matrix Inversion

The inverse for the full transformation matrix may be calculated from the following equation:

$$\boldsymbol{T}\boldsymbol{T}^{-1} = \boldsymbol{T}_1\boldsymbol{T}_2 = \begin{bmatrix} \boldsymbol{R}_1 & \boldsymbol{S}_1 \\ \mathbf{0} & 1 \end{bmatrix} \begin{bmatrix} \boldsymbol{R}_2 & \boldsymbol{S}_2 \\ \mathbf{0} & 1 \end{bmatrix} = \boldsymbol{I} \tag{2.60}$$

where $\boldsymbol{T}_1 = \boldsymbol{T}$ and $\boldsymbol{T}_2 = \boldsymbol{T}^{-1}$. Expanding this gives:

$$\boldsymbol{R}_1\boldsymbol{R}_2 + \boldsymbol{S}_1 \cdot \mathbf{0} = \boldsymbol{I} \tag{2.61}$$

$$\boldsymbol{R}_1\boldsymbol{S}_2 + \boldsymbol{S}_1 = \mathbf{0} \tag{2.62}$$

$$\mathbf{0} \cdot \boldsymbol{R}_2 + 1 \cdot \mathbf{0} = \mathbf{0} \tag{2.63}$$

$$\mathbf{0} \cdot \boldsymbol{S}_2 + 1 \cdot 1 = 1 \tag{2.64}$$

Eq. (2.61) gives:

$$\boldsymbol{R}_2 = \boldsymbol{R}_1^{-1} = \boldsymbol{R}_1^T$$

because $\boldsymbol{R}_1$ is an orthogonal matrix. Eq. (2.62) gives:

$$\boldsymbol{S}_2 = -\boldsymbol{R}_1^{-1}\boldsymbol{S}_1 = -\boldsymbol{R}_1^T\boldsymbol{S}_1$$

That is, if the transformation matrix is:

$$\boldsymbol{T} = \begin{bmatrix} \boldsymbol{R} & \boldsymbol{S} \\ \mathbf{0} & 1 \end{bmatrix}$$

the full inverse transformation matrix may be written:

$$\boldsymbol{T}^{-1} = \begin{bmatrix} \boldsymbol{R}^T & -\boldsymbol{R}^T\boldsymbol{S} \\ \mathbf{0} & 1 \end{bmatrix} \tag{2.65}$$

Example 2.21 : Invers transformation.

Use Eq. (2.65) to calculate the inverse transformation matrix for matrix $\boldsymbol{T}_2$ in Example 2.20.

$$\boldsymbol{R} = \begin{bmatrix} 0 & 0 & 1 \\ 0 & 1 & 0 \\ -1 & 0 & 0 \end{bmatrix}, \quad \boldsymbol{S} = \begin{bmatrix} 4 \\ 0 \\ 0 \end{bmatrix}$$

Then we have:

$$\boldsymbol{R}^T = \begin{bmatrix} 0 & 0 & -1 \\ 0 & 1 & 0 \\ 1 & 0 & 0 \end{bmatrix}$$

and

$$-\boldsymbol{R}^T\boldsymbol{S} = -\begin{bmatrix} 0 & 0 & -1 \\ 0 & 1 & 0 \\ 1 & 0 & 0 \end{bmatrix}\begin{bmatrix} 4 \\ 0 \\ 0 \end{bmatrix} = -\begin{bmatrix} 0 \\ 0 \\ 4 \end{bmatrix} = \begin{bmatrix} 0 \\ 0 \\ -4 \end{bmatrix}$$

The reduced transformation matrix $\boldsymbol{t}_2^{-1}$ is then

$$\boldsymbol{t}_2^{-1} = \begin{bmatrix} 0 & 0 & -1 & 0 \\ 0 & 1 & 0 & 0 \\ 1 & 0 & 0 & -4 \end{bmatrix}$$

From Example 2.20 the vector $\boldsymbol{v}_2 = (4, 0, -5)$ is a result of the transformation of vector $\boldsymbol{v}' = (5, 0, 0)$. Carry out the matrix-vector multiplication $\boldsymbol{t}_2^{-1}\boldsymbol{v}_2$, using reduced matrix-vector multiplication, Eq. (2.59):

$$\boldsymbol{v}' = \boldsymbol{t}_2^{-1}\boldsymbol{v}_2 = \begin{bmatrix} 0 & 0 & -1 \\ 0 & 1 & 0 \\ 1 & 0 & 0 \end{bmatrix}\begin{bmatrix} 4 \\ 0 \\ -5 \end{bmatrix} + \begin{bmatrix} 0 \\ 0 \\ -4 \end{bmatrix} = \begin{bmatrix} 5 \\ 0 \\ 0 \end{bmatrix}$$

That is, the inverse transformation transforms the point (4, 0, -5) back to (5, 0, 0) on the x-axis, see Example 2.20.

2.2.2 Orientation in Space

By a 3×3 direction cosine matrix an orientation in space is uniquely defined; refer to the development of the transformation matrix in the previous section. Orientation may also be specified by rotation angles about the axes in space, however, the orientation is only unique if the order of rotation is also specified. This is best demonstrated by an example.

Example 2.22 : Rotation of matchbox.

Demonstrate the result of rotating a matchbox 90° about the y- and z-axis, first in the order y and then z, and then in the order z and y.

As we see in Figs. 2.3 - 2.5 the order of rotation makes all the difference for the resulting position.

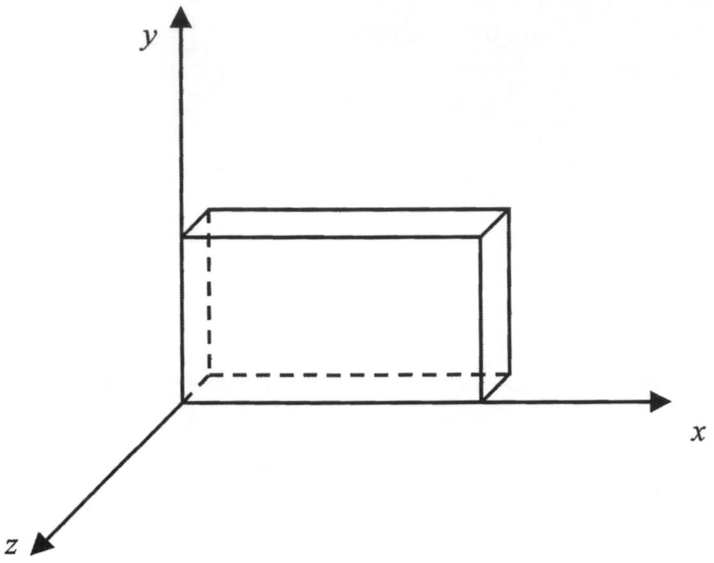

Figure 2.3: Initial position.

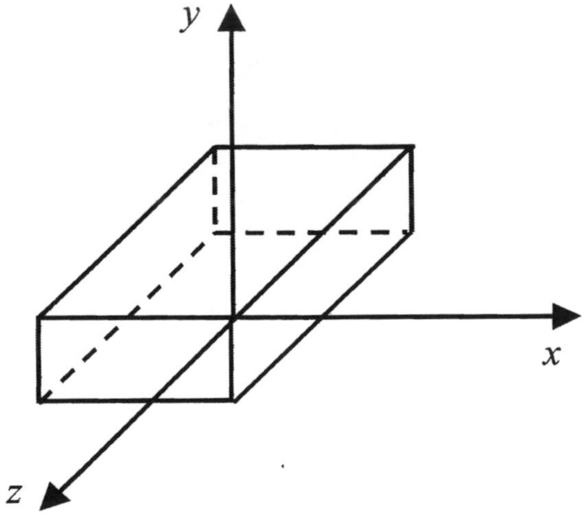

Figure 2.4: First rotation about y and then z.

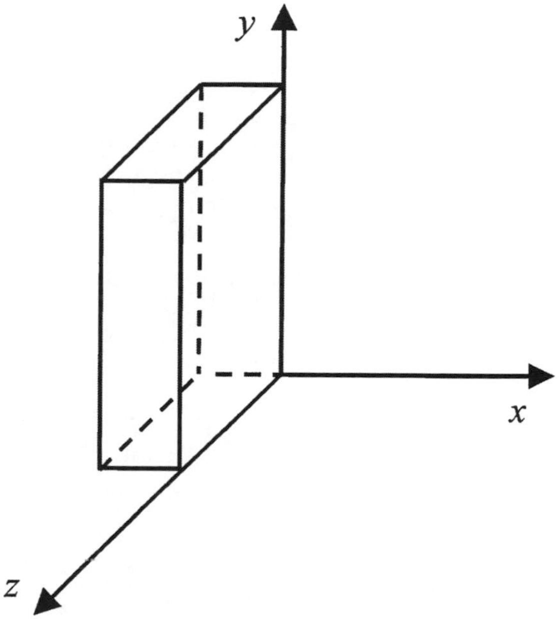

Figure 2.5: First rotation about z and then about y.

Example 2.23 : Transformation order.

Demonstrate, by matrix and vector operations, the transformations of Example 2.22. For 90° rotation about y- and z-axis, respectively, the transformation matrices may be written:

$$\boldsymbol{R}_y = \begin{bmatrix} 0 & 0 & 1 \\ 0 & 1 & 0 \\ -1 & 0 & 0 \end{bmatrix}, \quad \boldsymbol{R}_z = \begin{bmatrix} 0 & -1 & 0 \\ 1 & 0 & 0 \\ 0 & 0 & 1 \end{bmatrix}$$

First rotation about y and then about z, all referred to the original coordinate system, the transformation matrix is:

$$\boldsymbol{R}_1 = \boldsymbol{R}_z\boldsymbol{R}_y = \begin{bmatrix} 0 & -1 & 0 \\ 1 & 0 & 0 \\ 0 & 0 & 1 \end{bmatrix}\begin{bmatrix} 0 & 0 & 1 \\ 0 & 1 & 0 \\ -1 & 0 & 0 \end{bmatrix} = \begin{bmatrix} 0 & -1 & 0 \\ 0 & 0 & 1 \\ -1 & 0 & 0 \end{bmatrix}$$

Let us use this transformation to transform the point furthest away from the origin of the coordinate system. $\boldsymbol{v} = (10, 5, -2)$, refer to Fig. 2.3.

$$\boldsymbol{v}_1 = \boldsymbol{R}_1\boldsymbol{v} = \begin{bmatrix} 0 & -1 & 0 \\ 0 & 0 & 1 \\ -1 & 0 & 0 \end{bmatrix}\begin{bmatrix} 10 \\ 5 \\ -2 \end{bmatrix} = \begin{bmatrix} -5 \\ -2 \\ -10 \end{bmatrix}$$

Check this with Fig. 2.4. Then set up the transformation matrix for the case of first 90° rotation about z and then about y.

$$\boldsymbol{R}_2 = \boldsymbol{R}_y\boldsymbol{R}_z = \begin{bmatrix} 0 & 0 & 1 \\ 0 & 1 & 0 \\ -1 & 0 & 0 \end{bmatrix}\begin{bmatrix} 0 & -1 & 0 \\ 1 & 0 & 0 \\ 0 & 0 & 1 \end{bmatrix} = \begin{bmatrix} 0 & 0 & 1 \\ 1 & 0 & 0 \\ 0 & 1 & 0 \end{bmatrix}$$

$$\boldsymbol{v}_2 = \boldsymbol{R}_2\boldsymbol{v} = \begin{bmatrix} 0 & 0 & 1 \\ 1 & 0 & 0 \\ 0 & 1 & 0 \end{bmatrix}\begin{bmatrix} 10 \\ 5 \\ -2 \end{bmatrix} = \begin{bmatrix} -2 \\ 10 \\ 5 \end{bmatrix}$$

Check this with Fig. 2.5. The reader should transform other points for the matchbox in Figs. 2.3 - 2.5 to verify the transformed positions.

In opposite to Example 2.19, the rotations in Examples 2.22 and 2.23 are all referred to the original coordinate system. In computer graphics notation, this is called *picture mode* transformation as distinct from *space mode* transformation used in Example 2.19

As shown earlier, in space mode transformation, transformation matrices for subsequent transformations are postmultiplied into the current transformation matrix. For picture mode transformation, however, transformation matrices for subsequent transformations are premultiplied into the current transformation matrix.

The position and orientation of a point in space are uniquely specified by 6 degrees of freedom, 3 translations and 3 rotations, that is, 3 variables to specify a unique orientation. Direction cosines could not be used as variables, because they would introduce 9 variables for a unique specification of orientation. Rotation angles about 3 orthogonal axis in space, would introduce the 3 variables we need, but as shown above, these variables are dependent on the specified order of rotation, and are therefore not unique.

Small Rotations Commute

Rotation angles in space are not unique for *large rotations*. *Small rotation angles* - say infinitesimal - in space, however, are commutative, and could be used as variables for change in rotation.

Let us demonstrate the commutative behavior of small rotations in space. The small rotation angles about the x-, y- and z-axes are named $\Delta\alpha$, $\Delta\beta$ and $\Delta\gamma$, respectively. Setting sine to a small angle equal to the angle, cosine equal to 1 and product of small angles equal to zero, the transformation matrix from rotation about the x-, y- and z-axes in the named order is:

$$\begin{aligned} \boldsymbol{R}_z\boldsymbol{R}_y\boldsymbol{R}_x &= \begin{bmatrix} 1 & -\Delta\gamma & 0 \\ \Delta\gamma & 1 & 0 \\ 0 & 0 & 1 \end{bmatrix} \begin{bmatrix} 1 & 0 & \Delta\beta \\ 0 & 1 & 0 \\ -\Delta\beta & 0 & 1 \end{bmatrix} \begin{bmatrix} 1 & 0 & 0 \\ 0 & 1 & -\Delta\alpha \\ 0 & \Delta\alpha & 1 \end{bmatrix} \\ &= \begin{bmatrix} 1 & -\Delta\gamma & \Delta\beta \\ \Delta\gamma & 1 & -\Delta\alpha \\ -\Delta\beta & \Delta\alpha & 1 \end{bmatrix} \end{aligned}$$

or in the order of z-, y- and x-axis:

$$\begin{aligned} \boldsymbol{R}_x\boldsymbol{R}_y\boldsymbol{R}_z &= \begin{bmatrix} 1 & 0 & 0 \\ 0 & 1 & -\Delta\alpha \\ 0 & \Delta\alpha & 1 \end{bmatrix} \begin{bmatrix} 1 & 0 & \Delta\beta \\ 0 & 1 & 0 \\ -\Delta\beta & 0 & 1 \end{bmatrix} \begin{bmatrix} 1 & -\Delta\gamma & 0 \\ \Delta\gamma & 1 & 0 \\ 0 & 0 & 1 \end{bmatrix} \\ &= \begin{bmatrix} 1 & -\Delta\gamma & \Delta\beta \\ \Delta\gamma & 1 & -\Delta\alpha \\ -\Delta\beta & \Delta\alpha & 1 \end{bmatrix} \end{aligned}$$

That is, the resulting transformation for small rotations are uniquely defined by the 3 small rotation angles. This is shown for two different transformation orders above. Other combinations of order should be checked as an exercise for the reader. Hence, the transformation matrix for small rotations may be written (see also the skew-symmetric matrix, Eq. (2.20) and Example 2.9):

$$\Delta \boldsymbol{R} = \begin{bmatrix} 1 & -\Delta\gamma & \Delta\beta \\ \Delta\gamma & 1 & -\Delta\alpha \\ -\Delta\beta & \Delta\alpha & 1 \end{bmatrix} \tag{2.66}$$

A unique incremental vector for position and orientation in space may then be written:

$$\Delta \boldsymbol{v} = \begin{bmatrix} \Delta x \\ \Delta y \\ \Delta z \\ \Delta\alpha \\ \Delta\beta \\ \Delta\gamma \end{bmatrix} \tag{2.67}$$

Updating of Position and Orientation

A position and orientation in space is uniquely defined by the compact 3×4 transformation matrix, Eq. 2.54.

$$\boldsymbol{t} = \begin{bmatrix} R_{11} & R_{12} & R_{13} & X \\ R_{21} & R_{22} & R_{23} & Y \\ R_{31} & R_{32} & R_{33} & Z \end{bmatrix} \tag{2.68}$$

From calculated increments in the position and orientation variables, Eq. 2.67, a corresponding incremental transformation is set up:

$$\Delta \boldsymbol{t} = \begin{bmatrix} 1 & -\Delta\gamma & \Delta\beta & \Delta x \\ \Delta\gamma & 1 & -\Delta\alpha & \Delta y \\ -\Delta\beta & \Delta\alpha & 1 & \Delta z \end{bmatrix} \tag{2.69}$$

The new position is now calculated from the modified matrix product

$$\boldsymbol{t}_{n+1} = \boldsymbol{t}_n \Delta \boldsymbol{t}_n \tag{2.70}$$

where n is the increment number. Using the formula for modified matrix multiplication, Eq. (2.56), this may be written:

$$\boldsymbol{t}_{n+1} = [\boldsymbol{R}_n \Delta \boldsymbol{R}_n \quad \boldsymbol{S}_n + \boldsymbol{R}_n \Delta \boldsymbol{S}_n] \tag{2.71}$$

where

$\boldsymbol{R}_n$: direction cosines

$\Delta \boldsymbol{R}_n$: increment in direction cosines

$\boldsymbol{S}_n$: translation vector

$\Delta \boldsymbol{S}_n$: increment in translation vector

An Orientation in Space Definition

As discussed in detail above, large rotations about orthogonal axes in space are only unique if the order of rotation also is specified. We will now define an orientation in space by finite rotations about the orthogonal axes, and with the order of rotation specified, that is α, β and γ are defined as rotation angles about x-, y- and z-axes in that order. Having two directions in space uniquely defined by two direction cosine matrices defining the coordinate system xyz and $x^b y^b z^b$, calculate the rotation angles defined above relative the xyz coordinate system. This is done as follows; refer to Fig. 2.6:

- rotate the $x^b y^b x^b$ coordinate system about the z-axis, taking sign of rotation into account, until the x^b axis is parallel to the xz-plan. The rotation angle is $-\gamma$.

- transform the $x^b y^b z^b$ coordinate system with an angle of $-\gamma$ about the z-axis to position $x''y''z''$.

- rotate the $x''y''z''$ coordinate system about the y-axis, taking the sign of rotation into account, until the x'' is parallel to the x-axis. The rotation angle is $-\beta$.

- transform the $x''y''z''$ coordinate system with an angle of $-\beta$ about the y-axis to position $xy'z'$.

- rotate the $xy'z'$coordinate system about the x-axis, taking sign of rotation into account, until the z'-axis is parallel to the z-axis. The rotation angle is $-\alpha$.

The angles α β and γ may be calculated as follows. The direction cosine matrices for the xyz and $x^b y^b z^b$ coordinate systems, respectively, are written:

$$\boldsymbol{t} = [\boldsymbol{R}\ \boldsymbol{S}] \tag{2.72}$$

and

$$\boldsymbol{t}_b = [\boldsymbol{R}_b\ \boldsymbol{S}_b] \tag{2.73}$$

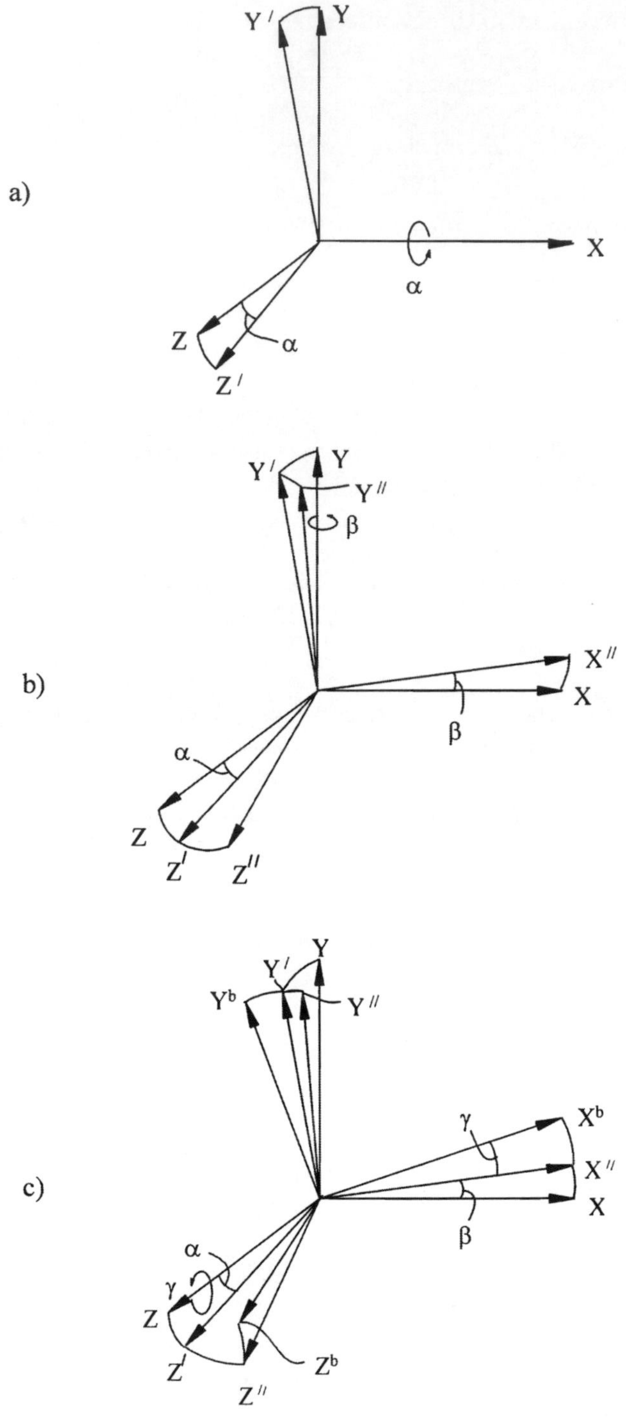

Figure 2.6: Rotation angles in space.

Calculate the relative transformations referred to the xyz coordinate system.

$$\boldsymbol{t}_c = \boldsymbol{t}^{-1}\boldsymbol{t}_b = \begin{bmatrix} R_{11} & R_{12} & R_{13} & S_1 \\ R_{21} & R_{22} & R_{23} & S_2 \\ R_{31} & R_{32} & R_{33} & S_3 \end{bmatrix} \tag{2.74}$$

The angle γ is now calculated from

$$-\gamma = -\arctan\left(R_{21}, R_{11}\right) \tag{2.75}$$

The angle γ about the z-axis is used to generate a transformation matrix $\boldsymbol{t}_z$ to the $x''y''z''$ position, see Eq. (2.47):

$$\boldsymbol{t}_z\left(-\gamma\right) = \boldsymbol{t}_z = \begin{bmatrix} \cos\left(-\gamma\right) & -\sin\left(-\gamma\right) & 0 & 0 \\ \sin\left(-\gamma\right) & \cos\left(-\gamma\right) & 0 & 0 \\ 0 & 0 & 1 & 0 \end{bmatrix} \tag{2.76}$$

The relative $x''y''z''$ transformation is calculated from:

$$\boldsymbol{t}_c'' = \boldsymbol{t}_c\boldsymbol{t}_z = \begin{bmatrix} R_{11}'' & R_{12}'' & R_{13}'' & S_1'' \\ R_{21}'' & R_{22}'' & R_{23}'' & S_2'' \\ R_{31}'' & R_{32}'' & R_{33}'' & S_3'' \end{bmatrix} \tag{2.77}$$

The angle β is now calculated from

$$-\beta = \arctan\left(R_{31}'', R_{11}''\right) \tag{2.78}$$

The angle β about the y-axis is used to generate a transformation matrix $\boldsymbol{t}_y$ to the $xy'z'$ position, refer to Eq. (2.49).

$$\boldsymbol{t}_y\left(-\beta\right) = \boldsymbol{t}_y = \begin{bmatrix} \cos\left(-\beta\right) & 0 & \sin\left(-\beta\right) & 0 \\ 0 & 1 & 0 & 0 \\ -\sin\left(-\beta\right) & 0 & \cos\left(-\beta\right) & 0 \end{bmatrix} \tag{2.79}$$

The relative $xy'z'$ transformation is calculated from

$$\boldsymbol{t}_c' = \boldsymbol{t}_c''\boldsymbol{t}_y = \begin{bmatrix} R_{11}' & R_{12}' & R_{13}' & S_1' \\ R_{21}' & R_{22}' & R_{23}' & S_2' \\ R_{31}' & R_{32}' & R_{33}' & S_3' \end{bmatrix} \tag{2.80}$$

The angle α is then calculated from

$$-\alpha = -\arctan\left(R_{32}', R_{22}'\right) \tag{2.81}$$

The relative transformation from the xyz to the $x^by^bz^b$ coordinate system may then be written

$$\boldsymbol{t}_{br} = \boldsymbol{t}_z\left(\gamma\right)\boldsymbol{t}_y\left(\beta\right)\boldsymbol{t}_x\left(\alpha\right) \tag{2.82}$$

For relative small rotations, say $5 - 10^o$, the errors introduced by using these rotation angles as variables are quite small. The same is true if two of the angles are small and the third is large.

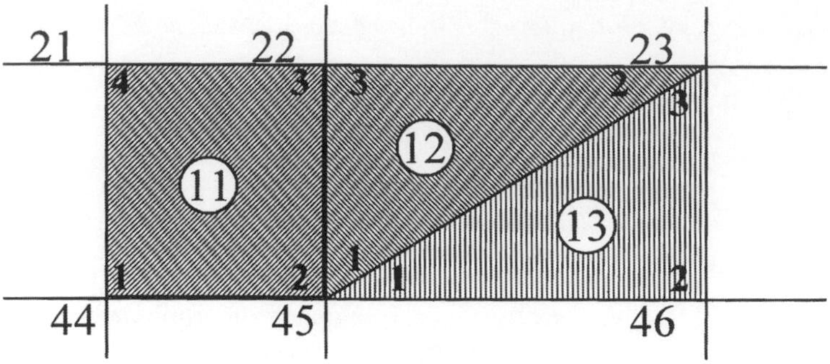

Figure 2.7: Mesh connectivity. Element numbers are circled, small numbers are local element node numbers and large numbers are system node numbers.

2.3 Finite Element Techniques

The *finite element method* is based on a subdivision of a geometric body into a number of sub-bodies, so-called finite elements, that is connected together by nodes into an element mesh. Examples of one-, two- and three-dimensional elements are beam, shell and volume elements, respectively. Fig. 2.8 shows some typical finite elements and their local node numbering (the text in this section is adopted from Bell, K. (1987)).

2.3.1 Element Connectivity

For definition of connectivity of the elements in a body, tables are set up coupling the local element node numbers mentioned above to a global node numbering of the overall body, refer to Fig. 2.7.

Each element has a connectivity vector called *"Matrix of Nodal Point Correspondence"*, (MNPC). MNPC for elements 11, 12 and 13 from Fig. 2.7 are, respectively (44, 45, 22, 21), (45, 23, 22) and (45, 46, 23). *"Matrix of MNPCs for all elements"* (MMNPC) contains the connectivity vectors MNPC for all elements of the model, listed consecutively in the matrix. The sequence of MNPCs in MMNPC corresponds to increasing element numbers.

As indicated in Fig. 2.7, the elements of the system model need not have the same number of nodes per element. In order to localize MNPC for a given element in MMNPC it is necessary to introduce the pointer vector called *"Matrix of Pointers to MNPC"* (MPMNPC). The length of MPMNPC is equal to the number of elements (NEL) in the model plus 1. If IEL is the element number, then MPMNPC(IEL) contains the index (pointer) in MMNPC where MNPC for that elements starts. Note that MPMNPC(NEL+1) con-

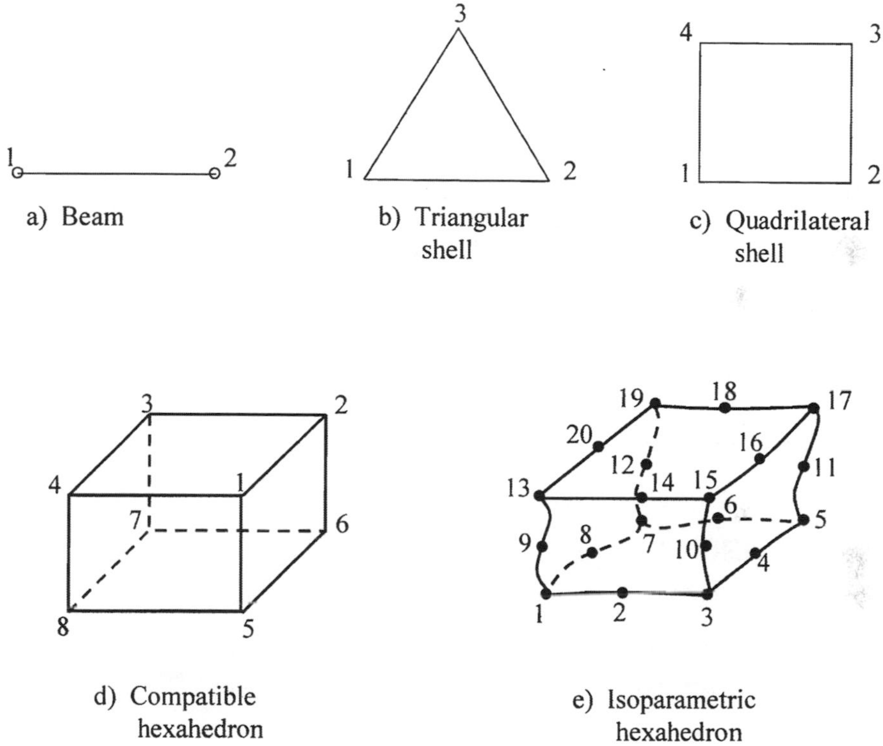

Figure 2.8: Finite elements.

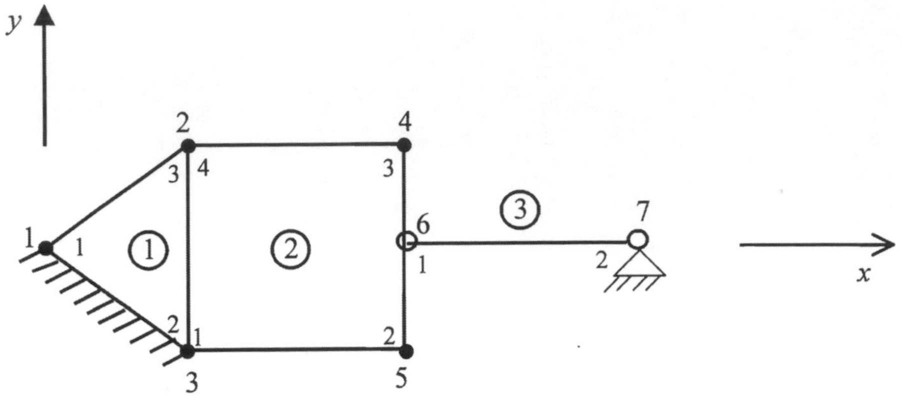

Figure 2.9: Finite element mesh.

tains the last index of MMNPC plus 1, that is, points to the first location succeeding MMNPC.

Example 2.24 : Element topology.

Set up the element topology for the mesh in Fig. 2.9 using matrices MPMNPC and MMNPC.

The matrices are as follows:

MPMNPC/1,4,8,10/

MMNPC/1,3,2;3,5,4,2;6,7/

2.3.2 Degrees of Freedom Status

All *degrees of freedom* (DOFs) of a system model must be assigned a *status code*. The status code 0, 1 and 2 are used, where 0 represents specified (eliminated) DOFs while 1 and 2 represents free DOFs, that is the unknowns in the model.

Only equations corresponding to free (unknown) DOFs are assembled, in other words, rows and columns corresponding to 0-status DOFs are not included in the system matrices.

The assembled system matrices correspond to a sequence of the unknown DOFs in which all 1-status DOFs appear before the 2-status DOFs. In substructure analysis, for instance, all internal or local DOFs are given status code 1, whereas all external or retained DOFs are assigned a 2-status. Note that if there is no need to rearrange the free DOFs (as in substructure analysis) all DOFs should be given the same statues code, 1 or 2.

"Matrix of Status Code" (MSC) is used for storing the status codes for the different DOFs of a system model, and the length of MSC is equal to the total number of DOFs (NDOF). In other words, MSC (IDOF) is the status code of DOF number IDOF. It is assumed that the DOFs are arranged in standard nodal point order (node by node).

The number of DOFs per node may vary from node to node. In order to localize node DOFs for individual nodes in MSC it is necessary to introduce the pointer vector called *"Matrix of Accumulated DOFs"* (MADOF). The length of MADOF is equal to the total number of nodes (NNOD) in the model plus 1. MADOF (NNOD + 1) contains the last index of MSC plus 1, that is, points to the first location succeeding MSC.

Example 2.25 : Accumulated DOFs.

Set up the matrix of accumulated DOFs (MADOF) taking into account that nodes 1-5 have two DOFs each while nodes 6 and 7 have three DOFs. Refer to Fig. 2.9 in Example 2.24.

The matrix is as follows:

MADOF/1,3,5,7,9,11,14,17/

From the matrices MSC and MADOF the parameters NDOF1, NDOF2 and NSPDOF are derived representing the number of *1-status* DOFs, the number of *2-status* DOFs and the number of *specified* (0-status) DOFs, respectively.

Referring to the total number of DOFs NDOF, as mentioned above, the following holds: NDOF = NDOF1 + NDOF2 + NSPDOF

It should also be noted that in standard nodal point order the DOFs are numbered consecutively from 1 to NDOF, starting with node no. 1, then node no. 2 and so on up to and including the last node NNOD.

2.3.3 Boundary Conditions

The boundary conditions are introduced through the specified DOFs mentioned above, and marked in the corresponding position in MSC by 0. Three types of boundary conditions for DOFs are defined that is named *suppressed, prescribed* and *dependent*, respectively. DOFs that is set equal to zero are called suppressed, DOFs given a fixed non-zero value are called prescribed, while DOFs that are expressed as predefined linear combination of a number ($\geq$1) of free DOFs are called dependent. In the following dependent DOFs will also be called *slave* DOFs while the free DOFs the slave refers to will be called *master* DOFs.

The types of boundary conditions mentioned above are all expressed by the general linear constraint equation.

$$r_i = r_{slave} = c_o + \sum_{j=1}^{P} c_j (r_{master})_j \tag{2.83}$$

where r_i refers to degree of freedom no. i, c_j $(j = 0, 1...P)$ are constants, and P are the number of masters for a slave DOF.

It is, however, convenient to distinguish between these three types of constraints, as they influence the system matrices in different ways.

Suppressed DOFs ($c_o = 0$, $P = 0$) represent the simplest type of boundary conditions. They are uniquely defined by the 0-status in MSC, and they do not influence the system matrices, except that the corresponding rows and columns are omitted in the system matrices.

Prescribed DOFs ($c_o = \delta$, $P = 0$) are as the suppressed DOFs eliminated from the system matrices, but they also introduce a load effect and thus produce modification to the load vectors.

For dependent DOFs ($P > 0$) the full linear constraint equation, Eq. (2.83), above apply.

In addition to the 0-status in MSC for prescribed and dependent DOFs, the c_o coefficient must be stored for the prescribed DOFs, and for dependent DOFs also slave and master node indexes and coefficients must be stored, refer to Eq. (2.83).

"Matrix of Constraint EQuation definition" (MCEQ) are of length P+1, and contains the DOF-number for the slave (ISL) and the masters (IM1, IM2, ... , IMP) in the above mentioned order.

"Table of Constraint Coefficients" (TCC) are also of length P+1, and contains the coefficients c_o, c_1, ..., c_p.

For prescribed DOFs P will be 0, and the matrices defined above will be of length 1. Specified DOFs not explicitly defined as prescribed or dependent in the matrices MCEQ and TCC are taken to be suppressed. A number of constraint equations (NCEQ) may be specified, and *"Matrix of MCEQs for all constraint equations"* (MMCEQ) and *"Table of TCC's for all constraint equations"* (TTCC) contains MCEQ and TCC, respectively, in the same order for all constraint equations defined in a model.

In order to localize the different MCEQ and TCC, in general of varying length, in MMCEQ and TTCC, *"Matrix of Pointers to MCEQ"* (MPMCEQ) is introduced. The length of MPMCEQ is NCEQ + 1 and MPMCEQ (NCEQ + 1) contains the last index of MMCEQ plus 1, respectively the last index of TTCC plus 1, that is, points to the first location succeeding MMCEQ, respectively TTCC.

Example 2.26 : Linear dependency.

Refering to Fig. 2.9 in Example 2.24. We assume that the relations between the deformations in nodes 4, 5 and 6 in x- and y-directions, respectively, can be expressed by the equations:

$$\begin{aligned} v_{6x} &= \frac{1}{2}v_{4x} + \frac{1}{2}v_{5x} \\ v_{6y} &= \frac{1}{2}v_{4y} + \frac{1}{2}v_{5y} \end{aligned}$$

Set up the matrices MSC, MPMCEQ, MMCEQ and TTCC.

Refering to Example 2.25, the degree of freedom status may be set up by inspection:

MSC/0,0;1,1;0,0;1,1;1,1;0,0,1;0,0,1/

Also by inspection the constraint matrices are:

MPMCEQ/1,4,7/

MMCEQ/11,7,9;12,8,10/

TTCC/0,0.5,0.5;0,0.5,0.5/

2.3.4 Element and System Matrices

Element Matrices

For each of the finite elements shown in Fig. 2.8 element matrices for stiffness, mass and possibly damping, may be generated. The element matrices are square and symmetric, and the dimension is dependent on the number of element DOFs. No distinction is made between *basic elements* and *super elements*. In both cases the square, symmetric element matrices must be available as full matrices.

The number of element nodes (NENOD) may vary from element to element, and the numbers of DOFs per node (NODDOF) may vary from node to node. This information is recorded in the control matrices MPMNPC, MMNPC and MADOF defined earlier.

System Matrices

The system model matrices for stiffness, mass and damping, respectively, are assembled by adding corresponding element matrices into the system matrices.

The order of DOFs in the system matrices are in general different from the standard nodal point order defined above, used for instance for MADOF and MSC. The system matrices are assembled to correspond to a rearranged order of the DOFs, where only free (unspecified) DOFs are included, and in such a way that all 1-status DOFs appear before the 2-status DOFs. The relative order of the rearranged DOFs within 1-status and 2-status DOFs, respectively, are, however, the same as for the standard nodal point order. The rearranged order of 1-status and 2-status DOFs separated and 0-status DOFs eliminated are called the equation order in contrast to the nodal point order. In other words, the first NDOF1 DOFs in equation order are 1-status DOFs while the following NDOF2 DOFs are 2-status DOFs. The dimension of the system matrices are equal to the Number of Equations (NEQ) at system level (NEQ = NDOF1 + NDOF2).

For a particular DOF, recognized by its DOF number in the nodal point ordering, the corresponding equation number may be found in the *"Matrix of EQuation Numbers"* (MEQN) with length NDOF. The content of MEQN are as follows

$$IEQ = MEQN(IDOF)$$

where

IEQ > 0: indicates that DOF number IDOF correspond to equation number IEQ

IEQ = 0: indicates that DOF number IDOF is suppressed

IEQ < 0: indicates that DOF number IDOF is constrained (prescribed or dependent), and |IEQ| is the pointer or index in MPMCEQ for this constrained DOF

MEQN is determined on the basis of the user-defined matrices MSC, MPMCEQ and MMCEQ described earlier.

Example 2.27 : Equation numbers.

Referring to Fig. 2.9 and Examples 2.24, 2.25 and 2.26. Set up the matrix of equation numbers (MEQN).

The matrix of eqation numbers are set up by inspection:

MEQN/0,0;1,2;0,0;3,4;5,6;-1,-2,7;0,0,8/

System Matrix Storage Format

The system matrices (stiffness, mass, damping) are also symmetric as are the elements matrices coming from the symmetric adding of element matrices into the system matrices. Because of usually very large systems matrices, the symmetry property is first utilized to let the elements below the diagonal be implied by the elements above the diagonal, and second the usual band property along the diagonal makes it possible to only store a band of the matrices that is wide enough to include all non-zero elements on and above the diagonal.

The storage scheme adopted for the system matrices is the so-called active column storage method, also known as the profile or *skyline* storage format.

The active columns of the upper triangular part of the matrix, each containing all elements from and including the first non-zero element to and including the diagonal element, are stored one after the other in a one-dimensional array as shown in Fig. 2.10. The location of each diagonal element within this array, defines the *"skyline"* and is recorded in a control matrix called *"Matrix of SKYline definition"* (MSKY). The non-zero elements in the square system matrix in Fig. 2.10 indicates the storage sequence of the system matrix in the one-dimensional system matrix array.

The last element of MSKY contains the length of the one-dimensional system matrix stored on skyline form. The dimension of MSKY is equal to Number of EQuations (NEQ = NDOF1 + NDOF2) defined earlier.

An arbitrary element located in row I and column J of a full square system matrix, on or above the diagonal (J $\geq$ I) but below the skyline, is located at address INDX in the skyline storage format, that is

$$SMFULL(I, J) = SMSKY(INDX)$$

where

INDX = MSKY (J) - J + I

The control matrix MSKY is determined on the basis of

- element-system connectivity: MMNPC and MPMNPC
- equation-DOF correspondence: MEQN and MADOF
- constraint equations: MMCEQ and MPMCEQ

Standardized control parameters are summarized in Table 2.1 and control matrices in Table 2.2.

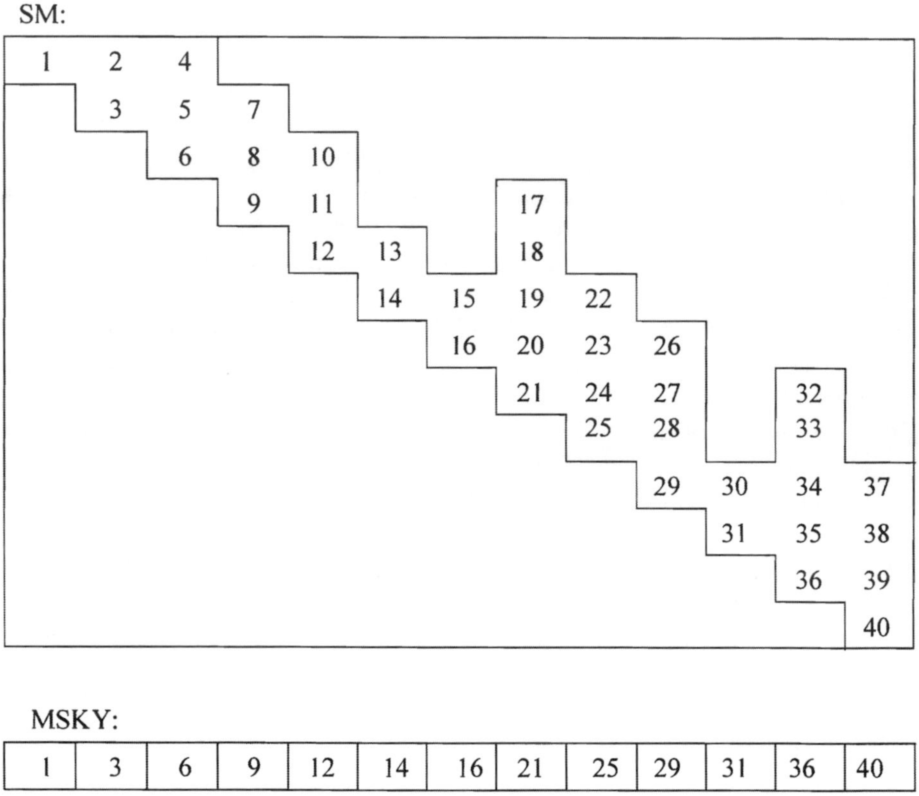

Figure 2.10: The system matrix (SM) on "skyline" form, and the MSKY vector that points to the bottom (diagonal) element of each stored column.

Table 2.1: Control Parameters.

NANOD:	Number of Active NODal points
NEL:	Number of ELements
NDOF:	Number of Degrees Of Freedom
NDOF1:	Number of 1-status DOFs
NDOF2:	Number of 2-status DOFs
NSPDOF:	Number of SPecified DOFs
NCEQ:	Number of Constraint EQuations
NSDOF:	Number of Suppressed DOFs
NPDOF:	Number of Prescribed DOFs
NDDOF:	Number of Dependent (slave) DOFs
NEQ:	Number of EQuations (NDOF1 + NDOF2)
NESKY:	Number of Elements in "SKYline array"
NLBW:	Number of elements in Largest BandWidth
NODLBW:	NODe number associated with NLBW
NMMNPC:	Number of elements in MMNPC
NMMCEQ:	Number of elements in MMCEQ

Table 2.2: Control Matrices.

MADOF:	Matrix of Accumulated DOFs
MNPC:	Matrix of Nodal Point Correspondence
MMNPC:	Matrix of MNPCs for all elements
MPMNPC:	Matrix of Pointers to MNPC in MMNPC
MSC:	Matrix of Status Codes
MCEQ:	Matrix of Constraint EQuation definition
MMCEQ:	Matrix of MCEQs for all constraint equations
TCC:	Table of Constraint Coefficients
TTCC:	Table of TCCs for all constraint equations
MPMCEQ:	Matrix of Pointers to MCEQ in MMCEQ and to TCC in TTCC
MEQN:	Matrix of EQuation Numbers
MSKY	Matrix of SKYline definition

Example 2.28 : Skyline matrix.

The matrix

$$M = \begin{bmatrix} 2 & 2 & 0 & 0 \\ 2 & 4 & 2 & 0 \\ 0 & 2 & 4 & 0 \\ 0 & 0 & 0 & 1 \end{bmatrix}$$

may be written on "skyline" form as:

MSKY/1,3,5,6/

SM/2;2,4;2,4;1/

2.4 Basic Numerical Algorithms

2.4.1 Equation Solution

For formulations based on the finite element method you always end up with a set of linear equations to solve. The number of equations to solve may be very large and in matrix form often written as:

$$\boldsymbol{A}\boldsymbol{x} = \boldsymbol{b} \tag{2.84}$$

where

$\boldsymbol{A}$: system matrix

$\boldsymbol{x}$: displacement vector

$\boldsymbol{b}$: load vector

For pure finite element formulations the system matrix is always symmetric and usually in band form, refer to the skyline storage format described in Section 2.3. Here we will show in principle the *Gauss's Elimination Method* on a full coefficient matrix. The matrix equation above may be written as

$$\begin{aligned} a_{11}x_1 + ... + a_{1n}x_n &= b_1 \\ a_{21}x_1 + ... + a_{2n}x_n &= b_2 \\ \cdots & \cdot \cdots \\ a_{m1}x_1 + ... + a_{mn}x_n &= b_m \end{aligned} \tag{2.85}$$

The Gauss algorithm will then be as follows:

First step

Elimination of x_1 from the second, third, ... up to m^{th} equation. We may assume that the order of the equations and the order of the unknowns in each equation are such that $a_{11} \neq 0$. The variable x_1 can then be eliminated from the second, third, ... up to the mth equation by subtracting $l_{21} = a_{21}/a_{11}$ times the first equation from the second equation, $l_{31} = a_{31}/a_{11}$ times the first equation from the third equation, etc. This gives a new system of equations of the form

$$\begin{aligned} a_{11}x_1 + a_{12}x_2 + ... + a_{1n}x_n &= b_1 \\ a_{22}^{(2)}x_2 + ... + a_{2n}^{(2)}x_n &= b_2^{(2)} \\ .... & \\ a_{m2}^{(2)}x_2 + ... + a_{mn}^{(2)}x_n &= b_m^{(2)} \end{aligned} \tag{2.86}$$

where the new coefficients are given from

$$a_{ij}^{(2)} = a_{ij} - l_{i1}a_{1j} \qquad b_i^{(2)} = b_i - l_{i1}b_1$$
$$i = 2, 3 \ldots, m$$

Any solution of Eq. (2.85) is a solution of Eq. (2.86) and conversely, because each equation of Eq. (2.86) was derived from two equations from Eq. (2.85) and, by reversing the process, Eq. (2.85) may be obtained from Eq. (2.86).

Second step

Elimination of x_2 from the third, fourth, ... up to m^{th} equation in Eq. (2.86). If the coefficients $a_{22}^{(2)}$, ..., $a_{mn}^{(2)}$ in Eq. (2.86) are not all zero, we may assume that the order of the equations and the unknowns is such that $a_{22}^{(2)} \neq 0$. Then we may eliminate x_2 from the third, fourth, ... , up to m^{th} equation of Eq. (2.86) by subtracting $l_{32} = a_{32}^{(2)}/a_{22}^{(2)}$ times the second equation from the third equation, $l_{42} = a_{42}^{(2)}/a_{22}^{(2)}$ times the second equation from the fourth equation, etc. The further steps are now obvious. In the third step we eliminate x_3, in the fourth step we eliminate x_4, etc.

The process will terminate only when no equations are left or when the coefficients of all unknowns in the remaining equations are zero. Introducing also $a_{ij}^{(1)} = a_{ij}$ we then have a system in the form

$$\begin{aligned} a_{11}^{(1)}x_1 + a_{12}^{(1)}x_2 + a_{13}^{(1)}x_3 + ... + a_{1n}^{(1)}x_n &= b_1^{(1)} \\ a_{22}^{(2)}x_2 + a_{23}^{(2)}x_3 + ... + a_{2n}^{(2)}x_n &= b_2^{(2)} \\ a_{33}^{(3)}x_3 + ... + a_{3n}^{(3)}x_n &= b_3^{(3)} \end{aligned} \tag{2.87}$$

$$\cdot$$
$$\cdot$$
$$\cdot$$

$$a_{rr}^{(r)}x_r + ... + a_{rn}^{(r)}x_n = b_r^{(r)}$$

where either $r = m$ or $r < m$. If $r < m$, the remaining equations have the form

$$0 = b_{r+1}^{(r)}, ..., \quad 0 = b_m^{(r)},$$

and the system has no solution, unless $b_{r+1}^{(r)} = 0, ..., b_m^{(r)} = 0$. If the system has a solution, we may obtain it by choosing values at pleasure for the unknowns $x_{r+1}, \ldots, x_m$, solving the last equation in Eq. (2.87) for x_r, the next to the last for x_{r-1}, and so on up to the first line.

In the important special case when $m = n = r$, Eq. (2.87) has triangular form and there is one, and only one, solution. For this special case and step k the linear set of equations may be written:

$$\sum_{j=k}^{n} a_{ij}^{(k)}x_j = b_i^{(k)}, \quad i = k, k+1, ...n,$$

where

$$\begin{aligned} a_{ij}^{(k)} &= a_{ij}^{(k-1)} - l_{i,k-1}a_{k-1,j}^{(k-1)} \\ j &= k, k+1, ...n \end{aligned}$$

and

$$b_i^{(k)} = b_i^{(k-1)} - l_{i,\,k-1}b_{k-1}^{(k-1)} \tag{2.88}$$

where

$$l_{i,k-1} = a_{i,k-1}^{(k-1)}/a_{k-1,k-1}^{(k-1)},$$
$$i = k, k+1, ..., n, \quad k = 2, 3, ..., n$$

with the initial values

$$a_{ij}^{(1)} = a_{ij}, \quad b_j^{(1)} = b_i$$

Example 2.29 : Gauss's elimination.

Solve the following set of equations:

$$\begin{aligned} 3x_1 + x_2 + 6x_3 &= 2 \\ 2x_1 + x_2 + 3x_3 &= 7 \\ x_1 + x_2 + x_3 &= 4 \end{aligned}$$

We make the following table of the coefficients that also include the right-hand side of the equation:

3	1	6	2
2	1	3	7
1	1	1	4

↓ First step

3	1	6	2
2/3	1/3	-1	17/3
1/3	2/3	-1	10/3

↓ Second step

3	1	6	2
2/3	1/3	-1	17/3
1/3	2	1	-8

The transformed set of equations may now be written:

$$\begin{aligned} 3x_1 + x_2 + 6x_3 &= 2 \\ \frac{1}{3}x_2 - x_3 &= 17/3 \\ x_3 &= -8 \end{aligned}$$

with the solution

$$x_3 = -8, \quad x_2 = -7 \quad \textit{and} \quad x_1 = 19$$

The Gauss's Elimination Method is in principle a transformation of a set of linear equations to triangular form. Let us assume that the coefficient matrix $\boldsymbol{A}$ may be factorized in a lower and upper triangular matrix called respectively $\boldsymbol{L}$ and $\boldsymbol{U}$ so that

$$\boldsymbol{Ax} = \boldsymbol{b} = \boldsymbol{LUx}$$

The transformed system is now split into two triangular set of equations:

$$\boldsymbol{Ly} = \boldsymbol{b}, \quad \boldsymbol{Ux} = \boldsymbol{y} \tag{2.89}$$

Theorem 2.1 : Matrix factorization.

Let $\boldsymbol{A}$ be a $n \times n$ matrix and assume that the equation

$$\boldsymbol{Ax} = \boldsymbol{b}$$

may be solved using the Gauss Elimination Method, then we have:
(i) $$\boldsymbol{A} = \boldsymbol{LU}$$
where

$$\boldsymbol{L} = \{l_{ik}\} = \begin{cases} 0 & , \quad i < k \\ 1 & , \quad i = k \\ a_{ik}^{(k)} / a_{kk}^{(k)} & , \quad i > k \end{cases}$$

and

$$\boldsymbol{U} = \{u_{ik}\} = \begin{cases} a_{ik}^{(i)} & , \quad i \leq k \\ 0 & , \quad i > k \end{cases}$$

(ii) $$det\boldsymbol{A} = a_{11}^{(1)} a_{22}^{(2)} ... a_{nn}^{(n)} \neq 0$$

Proof **:**

$$\begin{aligned} \text{Let} \quad \boldsymbol{LU} &= \{c_{ij}\} \quad \text{then} \\ c_{ij} &= \sum_{k=1}^{n} l_{ik} u_{kj} = \sum_{k=1}^{min(i,j)} l_{ik} a_{kj}^{(k)} \end{aligned}$$

We look at the two cases $i \leq j$ and $i > j$.
(1) $i \leq j$: From Eq. (2.88) we have

$$l_{i,k-1} a_{k-1,j}^{(k-1)} = a_{ij}^{(k-1)} - a_{ij}^{(k)}$$

and

$$\begin{aligned} c_{ij} &= \sum_{k=1}^{i-1} l_{ik} a_{kj}^{(k)} + a_{ij}^{(i)} \\ &= \sum_{k=1}^{i-1} \left(a_{ij}^{(k)} - a_{ij}^{(k+1)} \right) + a_{ij}^{(i)} = a_{ij}^{(1)} = a_{ij} \end{aligned}$$

(2) $i > j$: similarly

$$c_{ij} = \sum_{k=1}^{j} l_{ik} a_{kj}^{(k)} = a_{ij}^{(1)} - a_{ij}^{(j+1)} = a_{ij}$$

because

$$a_{ij}^{(j+1)} = 0$$

(i) of theorem 1 is now proved, and (ii) follows directly from

$$\begin{aligned} det\boldsymbol{A} &= det\boldsymbol{L} det\boldsymbol{U} = 1 \cdot det\boldsymbol{U} \\ &= a_{11}^{(1)} a_{22}^{(2)}, ... a_{nn}^{(n)} \\ & \quad QED \end{aligned}$$

Example 2.30 : Matrix factorizing.

Let us use ***LU*** factorizing on the linear set of equations from Example 2.29.

$$\boldsymbol{L} = \begin{bmatrix} 1 & 0 & 0 \\ a_{21}^{(1)}/a_{11}^{(1)} & 1 & 0 \\ a_{31}^{(1)}/a_{11}^{(1)} & a_{32}^{(2)}/a_{22}^{(2)} & 1 \end{bmatrix}$$

$$\tfrac{a_{32}^{(2)}}{a_{22}^{(2)}} = \tfrac{a_{32}^{(1)} - l_{31} \cdot a_{12}^{(1)}}{a_{22}^{(1)} - l_{21} \cdot a_{12}^{(1)}}$$

$$\begin{aligned} l_{31} &= a_{31}^{(1)}/a_{11}^{(1)} = 1/3 \\ l_{21} &= a_{21}^{(1)}/a_{11}^{(1)} = 2/3 \\ \frac{a_{32}^{(2)}}{a_{22}^{(2)}} &= \frac{1 - \frac{1}{3} \cdot 1}{1 - \frac{2}{3} \cdot 1} = \frac{2/3}{1/3} = 2 \end{aligned}$$

$$\boldsymbol{L} = \begin{bmatrix} 1 & 0 & 0 \\ 2/3 & 1 & 0 \\ 1/3 & 2 & 1 \end{bmatrix}$$

$$\boldsymbol{U} = \begin{bmatrix} a_{11}^{(1)} & a_{12}^{(1)} & a_{13}^{(1)} \\ 0 & a_{22}^{(2)} & a_{23}^{(2)} \\ 0 & 0 & a_{33}^{(3)} \end{bmatrix}$$

$$\begin{aligned} a_{22}^{(2)} &= a_{22}^{(1)} - l_{21} a_{12}^{(1)} = a_{22}^{(1)} - \tfrac{a_{21}^{(1)}}{a_{11}^{(1)}} \cdot a_{12}^{(1)} \\ &= 1 - \tfrac{2}{3} \cdot 1 = \tfrac{1}{3} \end{aligned}$$

$$a_{23}^{(2)} = a_{23}^{(1)} - l_{21} \cdot a_{13}^{(1)} = 3 - \frac{2}{3} \cdot 6 = -1$$

$$\begin{aligned} a_{33}^{(3)} &= a_{33}^{(2)} - l_{32} \cdot a_{23}^{(2)} = a_{33}^{(2)} - \tfrac{a_{32}^{(2)}}{a_{22}^{(2)}} \cdot a_{23}^{(2)} \\ &= a_{33}^{(1)} - l_{31} a_{13}^{(1)} - \tfrac{a_{32}^{(1)} - l_{31} a_{12}^{(1)}}{a_{22}^{(1)} - l_{21} a_{12}^{(1)}} \cdot \left(a_{23}^{(1)} - l_{21} \cdot a_{13}^{(1)} \right) \end{aligned}$$

$$\begin{aligned} a_{33}^{(3)} &= 1 - \tfrac{1}{3}6 - \tfrac{1-\frac{1}{3}\cdot 1}{1-\frac{2}{3}\cdot 1}\left(3 - \tfrac{2}{3}\cdot 6\right) \\ &= 1 - 2 - 2\cdot(-1) = 1 \end{aligned}$$

$$\boldsymbol{U} = \begin{bmatrix} 3 & 1 & 6 \\ 0 & 1/3 & -1 \\ 0 & 0 & 1 \end{bmatrix}$$

Using Eq. (2.89) we have

$$\boldsymbol{Ly} = \boldsymbol{b}$$

and vector $\boldsymbol{y}$ may now be calculated by what is called forward substitution

$$\begin{aligned} y_1 &= 2 \\ \frac{2}{3}y_1 + y_2 &= 7 \\ \frac{1}{3}y_1 + 2y_2 + y_3 &= 4 \end{aligned}$$

that gives:

$$y_1 = 2, \quad y_2 = \frac{17}{3}, \quad y_3 = -8$$

From Eq. (2.89) we also have

$$\boldsymbol{Ux} = \boldsymbol{y}$$

and by what is called backward substitution we can calculate vector $\boldsymbol{x}$

$$\begin{aligned} 3x_1 + x_2 + 6x_3 &= 2 \\ \frac{1}{3}x_2 - x_3 &= \frac{17}{3} \\ x_3 &= -8 \end{aligned}$$

that gives

$$x_3 = -8, \quad x_2 = -7, \quad x_1 = 19$$

This is the same as we had in Example 2.29 using Gauss elimination directly. Compare matrices $\boldsymbol{L}$, $\boldsymbol{U}$ and $\boldsymbol{y}$ with the last table in Example 2.29. That is, $\boldsymbol{LU}$-factorization is the same as Gauss Elimination, however, all coefficients are calculated in just one step. Another advantage from using the $\boldsymbol{LU}$-factorization is that as long as the coefficient matrix $\boldsymbol{A}$ is unchanged, the system of equations may be solved with new right-hand sides without any new triangularization, for instance, for new load cases or iteration sequences. For further studies on this subject refer to textbooks on linear algebra.

2.4.2 Eigenvalue Solution

Let $\boldsymbol{A} = \{a_{ij}\}$ be a given square n-rowed matrix and consider the vector equation

$$\boldsymbol{A}\boldsymbol{x} = \lambda \boldsymbol{x} \tag{2.90}$$

where λ is a number. It is clear that the zero vector $\boldsymbol{x} = 0$ is a solution for Eq. (2.90) for any value of λ. A value of λ for which Eq. (2.90) has a solution $\boldsymbol{x} \neq 0$ is called an *eigenvalue or characteristic value* of the matrix $\boldsymbol{A}$. The corresponding solution $\boldsymbol{x} \neq 0$ are called *eigenvectors or characteristic vectors* of $\boldsymbol{A}$ corresponding to eigenvalue λ. The set of eigenvalues is called the spectrum of $\boldsymbol{A}$.

The problem of determining the eigenvalues and eigenvectors of a matrix is called an eigenvalue problem. Problems of this type occur in connection with physical applications like dynamic simulations. Therefore we will show the fundamental ideas and concepts which are important in this field of mathematics. Over the last few decades various new methods for the approximate determination of eigenvalues have been developed and other methods which have been known for still a longer time have been put into a form which is suitable for computers.

Let us consider Eq. (2.90). If $\boldsymbol{x}$ is any vector, then the vectors $\boldsymbol{x}$ and $\boldsymbol{A}\boldsymbol{x}$ will, in general, be linear independent. If $\boldsymbol{x}$ is an eigenvector, then $\boldsymbol{x}$ and $\boldsymbol{A}\boldsymbol{x}$ are linearly dependent; corresponding components of $\boldsymbol{x}$ and $\boldsymbol{A}\boldsymbol{x}$ are then proportional, the factor of proportionality being the eigenvalue λ. We shall now demonstrate that any n-rowed square matrix has at least 1 and at most n distinct (real or complex) eigenvalues. For this purpose we write Eq. (2.90) at length:

$$\begin{aligned} a_{11}x_1 + \ldots + a_{1n}x_n &= \lambda x_1 \\ a_{21}x_1 + \ldots + a_{2n}x_n &= \lambda x_2 \\ \cdots\cdots\cdots&\cdots\cdots\cdots \\ a_{n1}x_1 + \ldots + a_{nn}x_n &= \lambda x_n \end{aligned} \tag{2.91}$$

or

$$\begin{aligned} (a_{11}-\lambda)\,x_1 + a_{12}x_2 + \ldots . + a_{1n}x_n &= 0 \\ a_{21}x_1 + (a_{22}-\lambda)\,x_2 + \ldots . + a_{2n}x_n &= 0 \\ \ldots\ldots\ldots\ldots\ldots\ldots\ldots\ldots\ldots\ldots\ldots & \\ a_{n1}x_1 + a_{n2}x_2 + \ldots . + (a_{nn}-\lambda)\,x_n &= 0 \end{aligned}$$

This homogeneous system of linear equations has a non-trivial solution if, and only if, the corresponding determinant of the coefficients is zero:

$$D(\lambda) = det\,(\boldsymbol{A} - \lambda \boldsymbol{I}) = \begin{vmatrix} a_{11}-\lambda & a_{12} & \ldots . & a_{1n} \\ a_{21} & a_{22}-\lambda & \ldots . & a_{2n} \\ . & . & \ldots . & . \\ a_{n1} & a_{n2} & \ldots . & a_{nn}-\lambda \end{vmatrix} = 0 \qquad (2.92)$$

$D(\lambda)$ is called the *characteristic determinant*, and Eq. (2.92) is called the *characteristic equation* corresponding to the matrix $\boldsymbol{A}$. By developing $D(\lambda)$ we obtain a polynomial of nth degree in λ. This is called the *characteristic polynomial* corresponding to $\boldsymbol{A}$. We have thus obtained the following important result.

Theorem 2.2 : Eigenvalue of matrix.

The eigenvalues of a square matrix $\boldsymbol{A}$ are the roots of the corresponding characteristic equation, Eq. (2.92).

An eigenvalue which is a root of mth order of the characteristic polynomial is called an eigenvalue of mth order of the corresponding matrix. Once the eigenvalues have been determined, the corresponding eigenvectors can be determined from the Eq. (2.91). Since the system is homogeneous, it is clear that if $\boldsymbol{x}$ is an eigenvector of $\boldsymbol{A}$, then $c\boldsymbol{x}$, where c is any constant, not zero, is also an eigenvector of $\boldsymbol{A}$ corresponding to the same eigenvalue.

Example 2.31 : Eigenvalues of matrix.

Determine the eigenvalues and eigenvectors of the matrix

$$\boldsymbol{A} = \begin{bmatrix} 5 & 4 \\ 1 & 2 \end{bmatrix}$$

The characteristic equation

$$D(\lambda) = \begin{vmatrix} 5-\lambda & 4 \\ 1 & 2-\lambda \end{vmatrix} = \lambda^2 - 7\lambda + 6 = 0$$

has the roots $\lambda_1 = 1$ and $\lambda_2 = 6$. For $\lambda = \lambda_1$ the Eq. (2.91) assumes the form

$$\begin{aligned} 4x_1 + 4x_2 &= 0 \\ x_1 + x_2 &= 0 \end{aligned}$$

Thus $x_1 = -x_2$, and

$$\boldsymbol{x}_1 = \begin{bmatrix} 1 \\ -1 \end{bmatrix}$$

is an eigenvector of $\boldsymbol{A}$ corresponding to the eigenvalue λ_1. In the same way we find that an eigenvector of $\boldsymbol{A}$ corresponding to λ_2 is

$$\boldsymbol{x}_2 = \begin{bmatrix} 4 \\ 1 \end{bmatrix}$$

$\boldsymbol{x}_1$ and $\boldsymbol{x}_2$ are linearly independent vectors.

2.4.3 Eigenvalues by Vector Iteration

Eigenvalue vector iteration for matrix $\boldsymbol{A}$ may be written:

$$\boldsymbol{z}_k = \boldsymbol{A}\boldsymbol{z}_{k-1} \quad k = 1, 2...... \tag{2.93}$$

where k is the iteration index. Starting this iteration with an arbitrary vector $\boldsymbol{z}_0$, $\boldsymbol{z}_k$ will converge to eigenvector $\boldsymbol{x}_n$ corresponding to the highest eigenvalue. This may be shown by

$$|\lambda_1| \leq |\lambda_2| \leq ....|\lambda_{n-1}| < |\lambda_n|$$

and that an arbitrary $\boldsymbol{z}_0$ may be expressed by

$$\boldsymbol{z}_0 = c_1\boldsymbol{x}_1 + c_2\boldsymbol{x}_2 + ... + c_n\boldsymbol{x}_n \tag{2.94}$$

where $\boldsymbol{x}_1$, $\boldsymbol{x}_2$ ... $\boldsymbol{x}_n$ are orthogonal eigenvectors to the matrix $\boldsymbol{A}$ and c_1, c_2, ... c_n are constants. Using Eqs. (2.93) and (2.94), first iteration may be written:

$$\boldsymbol{z}_1 = \boldsymbol{A}\boldsymbol{z}_o = c_1\lambda_1\boldsymbol{x}_1 + c_2\lambda_2\boldsymbol{x}_2 + ... + c_n\lambda_n\boldsymbol{x}_n$$

where λ_i and $\boldsymbol{x}_i$ are corresponding eigenvalues and eigenvectors. For iteration k the iteration equation gives

$$\boldsymbol{z}_k = \boldsymbol{A}\boldsymbol{z}_{k-1} = \boldsymbol{A}^k\boldsymbol{z}_0 = c_1\lambda_1^k\boldsymbol{x}_1 + ... + c_n\lambda_n^k\boldsymbol{x}_n$$

that may also be written

$$\boldsymbol{z}_k = \lambda_n^k \left\{ c_n \boldsymbol{x}_n + c_{n-1} \left(\frac{\lambda_{n-1}}{\lambda_n} \right)^k \boldsymbol{x}_{n-1} + \ldots + c_1 \left(\frac{\lambda_1}{\lambda_n} \right)^k \boldsymbol{x}_1 \right\}$$

This shows that all parts of the sum converges to zero for increasing k, except the part including $\boldsymbol{x}_n$. The relative size of λ_n and the other eigenvalues will effect the speed of the iteration convergence.

Example 2.32 : Vector iteration.

Use vector iteration to find the highest eigenvalue for the matrix of Example 2.31 choosing an "arbitrary" start vector to

$$\boldsymbol{z}_0 = \begin{bmatrix} 1 \\ 1 \end{bmatrix}$$

First iteration

$$\begin{bmatrix} 5 & 4 \\ 1 & 2 \end{bmatrix} \begin{bmatrix} 1 \\ 1 \end{bmatrix} = \begin{bmatrix} 9 \\ 3 \end{bmatrix} \sim \begin{bmatrix} 3 \\ 1 \end{bmatrix}$$

Second iteration

$$\begin{bmatrix} 5 & 4 \\ 1 & 2 \end{bmatrix} \begin{bmatrix} 3 \\ 1 \end{bmatrix} = \begin{bmatrix} 19 \\ 5 \end{bmatrix} \sim \begin{bmatrix} 3.8 \\ 1 \end{bmatrix}$$

Third iteration

$$\begin{bmatrix} 5 & 4 \\ 1 & 2 \end{bmatrix} \begin{bmatrix} 3.8 \\ 1 \end{bmatrix} = \begin{bmatrix} 23 \\ 5.8 \end{bmatrix} \sim \begin{bmatrix} 3.96 \\ 1 \end{bmatrix}$$

Forth iteration

$$\begin{bmatrix} 5 & 4 \\ 1 & 2 \end{bmatrix} \begin{bmatrix} 3.96 \\ 1 \end{bmatrix} = \begin{bmatrix} 23.82 \\ 5.96 \end{bmatrix} \sim \begin{bmatrix} 3.99 \\ 1 \end{bmatrix}$$

4 iterations give eigenvalue $\lambda_2 = 5.96$ and $\boldsymbol{x}_2 = \begin{bmatrix} 3.99 \\ 1 \end{bmatrix}$. Compare with Example 2.31.

2.4.4 Eigenvalue by Inverse Iteration

Usually it is only one or a few of the lowest eigenfrequencies that are of interest for the designer of a physical system. To achieve this we can formulate the inverse eigenvalue problem:

$$\boldsymbol{A}^{-1}\boldsymbol{x} = \frac{1}{\lambda}\boldsymbol{x} \tag{2.95}$$

The highest eigenvalue of $\boldsymbol{A}^{-1}$ is the inverse of the lowest eigenvalue for $\boldsymbol{A}$. To avoid inversion of the matrix $\boldsymbol{A}$ Eq. (2.93) may be rewritten as follows

$$\boldsymbol{A}\boldsymbol{z}_k = \boldsymbol{z}_{k-1} \tag{2.96}$$

and the iteration is now carried out by solving Eq. (2.96) for $\boldsymbol{z}_k$ repeatedly. As above for vector iteration, $\boldsymbol{z}_0$ may be selected arbitrary, however, a selection of $\boldsymbol{z}_0$ that in a better way reflects the actual eigenvector will speed up the convergence. The iteration will converge to the lowest eigenvalue and corresponding eigenvector.

Example 2.33 : Inverse iteration.

In this example we will use inverse iteration to find the lowest eigenvalue for the matrix in Example 2.31. We will first use $\boldsymbol{LU}$ factorizing for matrix $\boldsymbol{A}$.

$$\boldsymbol{A} = \begin{bmatrix} 5 & 4 \\ 1 & 2 \end{bmatrix} = \boldsymbol{LU}$$

where

$$\boldsymbol{L} = \begin{bmatrix} 1 & 0 \\ a_{21}^{(1)}/a_{11}^{(1)} & 1 \end{bmatrix} = \begin{bmatrix} 1 & 0 \\ 1/5 & 1 \end{bmatrix}$$

and

$$\boldsymbol{U} = \begin{bmatrix} a_{11}^{(1)} & a_{12}^{(1)} \\ 0 & a_{22}^{(2)} \end{bmatrix} = \begin{bmatrix} 5 & 4 \\ 0 & \frac{6}{5} \end{bmatrix}$$

where

$$a_{22}^{(2)} = a_{22}^{(1)} - \frac{a_{21}^{(1)}}{a_{11}^{(1)}} a_{12}^{(1)} - 2 - \frac{1}{5} \cdot 4 = \frac{6}{5}$$

Choosing the same "arbitrary" $\boldsymbol{z}_0$ as for Example 2.32. The iteration then proceeds as follows:

First iteration

$$\boldsymbol{L}\boldsymbol{y}_1 = \begin{bmatrix} 1 & 0 \\ \frac{1}{5} & 1 \end{bmatrix} \begin{bmatrix} y_1 \\ y_2 \end{bmatrix}_1 = \begin{bmatrix} 1 \\ 1 \end{bmatrix}_0 = \boldsymbol{z}_0$$

$$y_1 = 1, \quad y_2 = 1 - \tfrac{1}{5} = \tfrac{4}{5}$$

and

$$\boldsymbol{U}\boldsymbol{z}_1 = \begin{bmatrix} 5 & 4 \\ 0 & \frac{6}{5} \end{bmatrix} \begin{bmatrix} z_1 \\ z_2 \end{bmatrix}_1 = \begin{bmatrix} 1 \\ 4/5 \end{bmatrix}_1 = \boldsymbol{y}_1$$

$$z_2 = \tfrac{4}{6} = \tfrac{2}{3}, \quad z_1 = \tfrac{1}{5} - \tfrac{8}{5 \cdot 3} = -\tfrac{1}{3}$$

$$\boldsymbol{z}_1 = \begin{bmatrix} -\frac{1}{3} \\ \frac{2}{3} \end{bmatrix} \sim \begin{bmatrix} -\frac{1}{2} \\ 1 \end{bmatrix}$$

Second iteration

$$\boldsymbol{L}\boldsymbol{y}_2 = \begin{bmatrix} 1 & 0 \\ \frac{1}{5} & 1 \end{bmatrix} \begin{bmatrix} y_1 \\ y_2 \end{bmatrix}_2 = \begin{bmatrix} -\frac{1}{2} \\ 1 \end{bmatrix}_1 = \boldsymbol{z}_1$$

$$y_1 = -\tfrac{1}{2}, \quad y_2 = 1 + \tfrac{1}{10} = \tfrac{11}{10}$$

and

$$\boldsymbol{U}\boldsymbol{z}_2 = \begin{bmatrix} 5 & 4 \\ 0 & \frac{6}{5} \end{bmatrix} \begin{bmatrix} z_1 \\ z_2 \end{bmatrix}_2 = \begin{bmatrix} -1/2 \\ 11/10 \end{bmatrix}_2 = \boldsymbol{y}_2$$

$$z_2 = \tfrac{11}{10} \cdot \tfrac{5}{6} = \tfrac{11}{12}, \quad z_1 = -\tfrac{1}{2} - \tfrac{11}{3} = -\tfrac{5}{6}$$

$$\boldsymbol{z}_2 = \begin{bmatrix} -\frac{5}{6} \\ \frac{11}{12} \end{bmatrix} \sim \begin{bmatrix} -\frac{10}{11} \\ 1 \end{bmatrix}$$

Third iteration

$$\boldsymbol{L}\boldsymbol{y}_3 = \begin{bmatrix} 1 & 0 \\ \frac{1}{5} & 1 \end{bmatrix} \begin{bmatrix} y_1 \\ y_2 \end{bmatrix}_3 = \begin{bmatrix} -\frac{10}{11} \\ 1 \end{bmatrix}_2 = \boldsymbol{z}_2$$

$$y_1 = -\tfrac{10}{11}, \quad y_2 = \tfrac{13}{11}$$

and

$$\boldsymbol{U}\boldsymbol{z}_3 = \begin{bmatrix} 5 & 4 \\ 0 & \frac{6}{5} \end{bmatrix} \begin{bmatrix} z_1 \\ z_2 \end{bmatrix}_3 = \begin{bmatrix} -\frac{10}{11} \\ \frac{13}{11} \end{bmatrix}_3 = \boldsymbol{y}_3$$

$$z_2 = \tfrac{13}{11} \cdot \tfrac{5}{6} = 0.985, \quad z_1 = -0.97$$

$$\boldsymbol{z}_3 = \begin{bmatrix} -0.97 \\ 0.985 \end{bmatrix} \sim \begin{bmatrix} -0.985 \\ 1 \end{bmatrix}$$

After 3. iterations the eigenvalue $\lambda_1 = 1/0,985$ and the corresponding eigenvector

$$\boldsymbol{x}_1 = \begin{bmatrix} -.985 \\ 1 \end{bmatrix} \sim \begin{bmatrix} .985 \\ -1 \end{bmatrix}$$

Compare this with the analytical result of Example 2.31.

2.4.5 The General Eigenvalue Problem

The equation for the general eigenvalue problem may be written

$$(\boldsymbol{A} - \lambda \boldsymbol{B})\,\boldsymbol{x} = \boldsymbol{0} \tag{2.97}$$

where $\boldsymbol{A}$ and $\boldsymbol{B}$ are matrices. For dynamics in mechanical systems this equation may be written

$$(\boldsymbol{K} - \lambda \boldsymbol{M})\,\boldsymbol{x} = \boldsymbol{0} \tag{2.98}$$

where $\boldsymbol{K}$ is the stiffness matrix and $\boldsymbol{M}$ is the mass matrix. The iteration equation may now be written

$$\boldsymbol{K}\boldsymbol{z}_k = \boldsymbol{M}\boldsymbol{z}_{k-1} \to \omega^2 \boldsymbol{M}\boldsymbol{z}_k \tag{2.99}$$

where ω^2 is the lowest eigenvalue. From this we have

$$\omega^2 \boldsymbol{z}_k \approx \boldsymbol{z}_{k-1}$$

If the matrix $\boldsymbol{K}$ is invertible, the inverse iteration algorithm may be written as follows

$$\boldsymbol{K} = \boldsymbol{LU} \tag{2.100}$$

and Eq. (2.99) may be written

$$\boldsymbol{K z}_k = \boldsymbol{LU z}_k = \boldsymbol{L y}_k = \boldsymbol{M z}_{k-1}$$

where

$$\boldsymbol{y}_k = \boldsymbol{U z}_k$$

$\boldsymbol{y}_k$ is now found by forward substitution using

$$\boldsymbol{L y}_k = \boldsymbol{M z}_{k-1} \tag{2.101}$$

and

$$\boldsymbol{y}_k = \boldsymbol{L}^{-1}\boldsymbol{M z}_{k-1}$$

$\boldsymbol{z}_k$ is found by backward substitution using

$$\boldsymbol{U z}_k = \boldsymbol{y}_k \tag{2.102}$$

and

$$\boldsymbol{z}_k = \boldsymbol{U}^{-1}\boldsymbol{y}_k = \boldsymbol{K}^{-1}\boldsymbol{M z}_{k-1}$$

The matrix $\boldsymbol{K}$ is constant and is factorized only once at the start of the iteration.

The iteration converges toward the eigenvector, corresponding to the lowest eigenvalue that may be found from

$$\omega^2 = \lambda_R = \frac{\boldsymbol{z}_k^T \boldsymbol{M z}_{k-1}}{\boldsymbol{z}_k^T \boldsymbol{M z}_k} = \frac{\boldsymbol{z}_k^T \boldsymbol{K z}_k}{\boldsymbol{z}_k^T \boldsymbol{M z}_k} \tag{2.103}$$

This is called the Rayleigh ratio.

2.4.6 Inverse Iteration with Shift

To improve the convergence of the iteration and find more eigenvalues, iteration with shift may be used. The eigenvalue problem may then be written

$$\left((\boldsymbol{K} - \mu\boldsymbol{M}) - \overline{\lambda}\boldsymbol{M}\right)\boldsymbol{x} = \boldsymbol{0} \tag{2.104}$$

where $\overline{\lambda}_i$ are modified eigenvalues given by the relation

$$\overline{\lambda}_i = \lambda_i - \mu$$

where λ_i is the actual eigenvalues and μ the shift value. The eigenvectors will be the same as for the original eigenvalue problem. For iteration with shift the iteration equation may be written

$$(\boldsymbol{K} - \mu\boldsymbol{M})\,\boldsymbol{z}_k = \boldsymbol{M}\boldsymbol{z}_{k-1} \tag{2.105}$$

Matrix $\boldsymbol{K} - \mu\boldsymbol{M}$ must be invertible and are factorized as shown for matrix $\boldsymbol{K}$ earlier. The factorization must be repeated each time the shift value μ is changed in the search for new eigenvectors and corresponding eigenvalues. $\boldsymbol{z}_k$ converges towards the eigenvector that has corresponding eigenvalue closest to μ, that is the eigenvalue $\overline{\lambda}_i\,(= \lambda_i - \mu)$ with the smallest absolute value.

For all these eigenvalue iteration algorithms, the iteration vector $\boldsymbol{z}_k$ must in the general case be scaled after each iteration to avoid overflow in the computations. This algorithm is well suited for computer implementation. A number of eigenvalue algorithms are based on similar iteration schemes. For further studies refer to separate textbooks on this subject.

Chapter 3

Mechanism Modeling

3.1 Mechanism Initial Position

3.1.1 Joint and link coordinate systems

The objective for mechanism modeling is to build an assembled position of a mechanism in space where the different links are coupled through joints of various types. For kinematic analysis of a mechanism, only the orientation and position of the joints are of importance, while the link shapes are unimportant except for the relative position of the joints. For dynamic analysis, especially the flexible body dynamics, the geometry and properties of the links are vital for the simulation. However, simplified geometry is often used for initial dynamic simulation as for instance in the feasibility studies of a proposed design.

One typical situation is that the joint position and orientation are known for one assembled position of the mechanism for instance from a mechanism synthesis program or a CAD system. In this global position of the mechanism, each joint will contain two or more coordinate systems attached to the two links connected through each joint and with a direction according to the joint axis. For a revolute joint between links A and B the joint coordinate systems may have a design position as shown in Fig. 3.1.

For the initial position of the mechanism all joint coordinate systems are positioned relative to a global coordinate system. Fig. 3.2 is an example of an initial design position of a four-bar mechanism specified by the coordinate systems for the four revolute joints. Here the coordinate axis within each link is chosen parallel, that is Y_{2B} and Y_{3B} are parallel as are X_{2B} and X_{3B} and so on. In addition to the joint coordinate systems on a link, the link will also have a link coordinate system. The link coordinate system may coincide with one of the joint coordinate systems on the link or be a separate coordinate

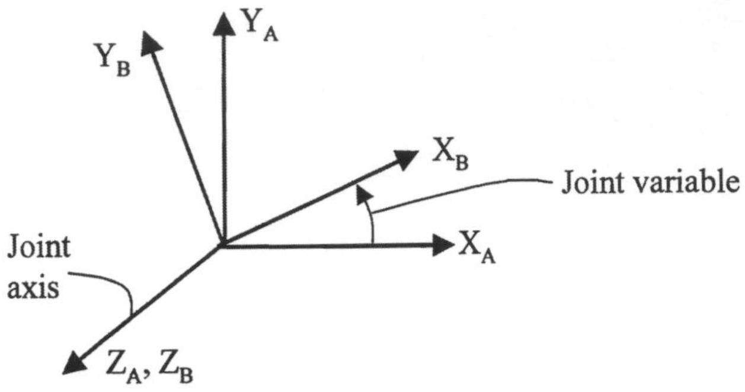

Figure 3.1: Joint coordinate system for revolute joint.

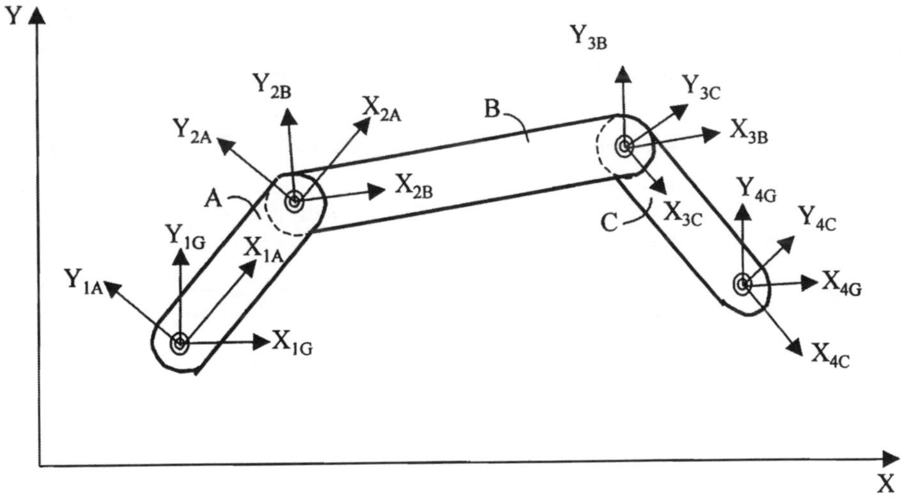

Figure 3.2: Joint position and orientation for the initial position of a four-bar mechanism with joint numbers 1-4 and links, A, B, C and G (ground).

system.

One way of specifying the joint and link coordinate systems can be to define the position and direction of the joint, respective, link coordinate systems by positioning three points for each coordinate system relative the global coordinate system. Typically the first point ($\boldsymbol{P}_1$) will be put in the origin for the coordinate system to be positioned, the second ($\boldsymbol{P}_2$) put on the positive x-axis and the third ($\boldsymbol{P}_3$) in the xy-plane in positive y-direction. The transformation between the global and this actual coordinate system are then calculated as follows:

1. Forming the two vectors $\boldsymbol{v}_x$ and $\boldsymbol{v}_A$ with direction from $\boldsymbol{P}_1$ to $\boldsymbol{P}_2$ and $\boldsymbol{P}_1$ to $\boldsymbol{P}_3$, respectively, where $\boldsymbol{e}_x = \boldsymbol{v}_x / \|\boldsymbol{v}_x\|$
2. Calculate the unit vector $\boldsymbol{e}_z$ from the cross product

$$\boldsymbol{v}_z = (\boldsymbol{v}_x \times \boldsymbol{v}_A) \tag{3.1}$$

where

$$\boldsymbol{e}_z = \boldsymbol{v}_z / \|\boldsymbol{v}_z\| \tag{3.2}$$

3. Calculate the unit vector $\boldsymbol{e}_y$ from $\boldsymbol{e}_y = (\boldsymbol{e}_z \times \boldsymbol{e}_x)$
4. If the coordinates for the first point are represented by $\boldsymbol{P}_1$, the 3×4 transformation matrix $\boldsymbol{t}$ (Eq. 2.54) for this coordinate system relative the global coordinate system is:

$$\boldsymbol{t} = [\boldsymbol{e}_x \; \boldsymbol{e}_y \; \boldsymbol{e}_z \; \boldsymbol{P}_1] \tag{3.3}$$

Example 3.1 : Transformation of links.

Calculate a 3×4 transformation matrix for a joint coordinate system relative a global system when the origin ($\boldsymbol{P}_1$) is positioned in (3, 1, 0), the point on the x-axis ($\boldsymbol{P}_2$) is positioned in (3, 4, 0) and the third point ($\boldsymbol{P}_3$) in (0, 0, 0).

Step 1:

$$\boldsymbol{e}_x = \frac{1}{3}\begin{bmatrix} 0 \\ 3 \\ 0 \end{bmatrix} = \begin{bmatrix} 0 \\ 1 \\ 0 \end{bmatrix} \qquad \boldsymbol{v}_A = \begin{bmatrix} -3 \\ -1 \\ 0 \end{bmatrix}$$

Step 2:

$$\boldsymbol{v}_z = (\boldsymbol{e}_x \times \boldsymbol{v}_A)$$

and

$$\boldsymbol{e}_z = \boldsymbol{v}_z / \|\boldsymbol{v}_z\|$$

Using Eq. (2.42) we have

$$\begin{aligned} v_{z1} &= 1 \cdot 0 - (-1) \cdot 0 = 0 \\ v_{z2} &= 0 \cdot (-3) - 0 \cdot 0 = 0 \\ v_{z3} &= 0 \cdot (-1) - (-3) \cdot 1 = 3 \\ \boldsymbol{v}_z &= \begin{bmatrix} 0 \\ 0 \\ 3 \end{bmatrix} \quad \boldsymbol{e}_z = \begin{bmatrix} 0 \\ 0 \\ 1 \end{bmatrix} \end{aligned}$$

Step 3:

$$\boldsymbol{e}_y = (\boldsymbol{e}_z \times \boldsymbol{e}_x)$$

Using Eq. (2.42) again we have

$$\begin{aligned} e_{y1} &= 0 \cdot 0 - 1 \cdot 1 = -1 \\ e_{y2} &= 1 \cdot 0 - 0 \cdot 0 = 0 \\ e_{y3} &= 0 \cdot 1 - 0 \cdot 0 = 0 \\ \boldsymbol{e}_y &= \begin{bmatrix} -1 \\ 0 \\ 0 \end{bmatrix} \end{aligned}$$

Step 4:

The 3×4 transformation matrix for the defined coordinate system is then

$$\boldsymbol{t} = [\boldsymbol{e}_x \; \boldsymbol{e}_y \; \boldsymbol{e}_z \; \boldsymbol{P}_1] = \begin{bmatrix} 0 & -1 & 0 & 3 \\ 1 & 0 & 0 & 1 \\ 0 & 0 & 1 & 0 \end{bmatrix}$$

referred to the global coordinate system.

In this way the transformation matrices for all joint and link coordinate systems are calculated for the initial position of the assembled mechanism. For rigid body analysis, the joint coordinate systems for each link are constant and fixed relative to the link coordinate system during mechanism motion.

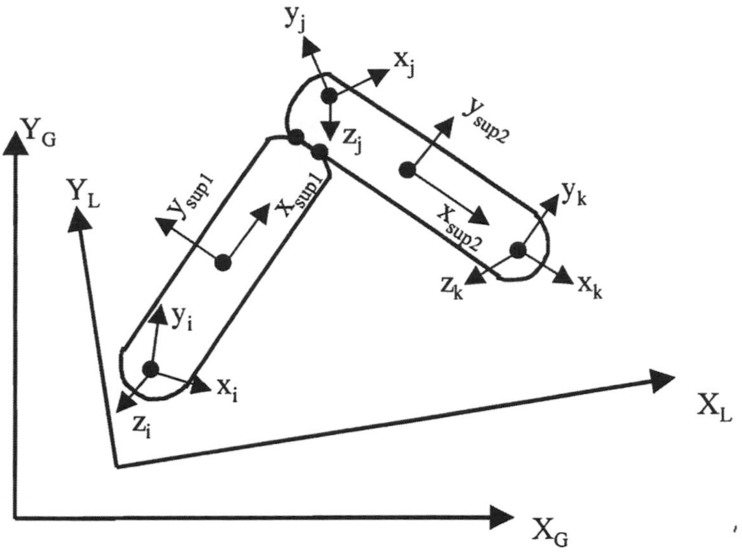

Figure 3.3: Super element and joint coordinate system for one link of the mechansim.

3.1.2 Super elements

For rigid body analysis the joint and link coordinate systems defined above are sufficient to define the mechanism unambiguously with its rigid body degrees of freedom. In the case of flexible body analysis, each link is built from one or more finite element models called super elements. Each super element is referred to its own coordinate system that is positioned relative the link coordinate system. The relative transformation between the link and the super element coordinate systems may be calculated in a similar way as above and is called $\boldsymbol{t}_{SL}$.

Let us now consider one link composed of two super elements and with three joints, see Fig. 3.3. The transformation for the link coordinate system in Fig. 3.3 is given relative the global system and named $\boldsymbol{t}_{LG}$. The transformations for the super element coordinates systems are given relative to the link system and named $\boldsymbol{t}_{SL}$.

The transformation for the super elements relative the global coordinate system named $\boldsymbol{t}_{SG}$ are calculated from

$$\boldsymbol{t}_{SG} = \boldsymbol{t}_{LG} \cdot \boldsymbol{t}_{SL} \tag{3.4}$$

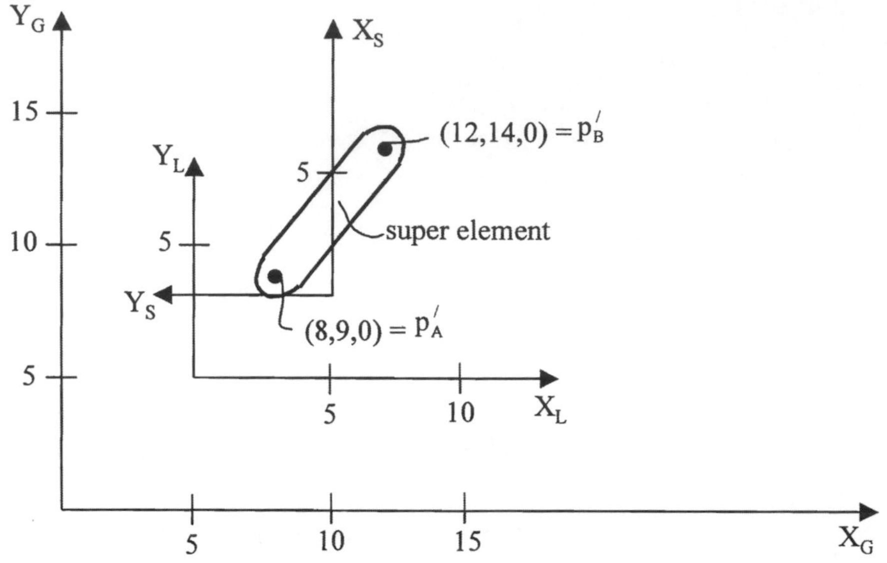

Figure 3.4: Link and super element coordinate systems.

This transformation may be used to position a super element from its position relative the link coordinate system to a position relative the global coordinate system.

When positioning the joints and links with their global position and orientation, as we proposed here, the transformation matrix $\boldsymbol{t}_{SG}$ from Eq. (3.4) will be used to transform each super element back to its local coordinate system to have exact position for the external nodes when modeling the mesh for the corresponding substructure. For this we need the inverse of $\boldsymbol{t}_{SG}$.

Example 3.2 : Inverse link transformation.

For transforming the super element back to the local coordinate system using matrix multiplication, refer to Fig. 3.4.

The transformation matrices $\boldsymbol{t}_{LG}$ and $\boldsymbol{t}_{SL}$ for the link and the super element, respectively, may be calculated from three positioned points for each as in the Example 3.1. However, for this simple geometry these matrices may be set up directly by inspection. For the link coordinate system the axis are parallel to the corresponding global axis, only with a translation. The matrix is therefore as follows, refer to Eq. (3.3):

$$t_{LG} = \begin{bmatrix} 1 & 0 & 0 & 5 \\ 0 & 1 & 0 & 5 \\ 0 & 0 & 1 & 0 \end{bmatrix}$$

The coordinate system for the super element relative the link coordinate system has its x-axis parallel to the link y-axis, the y-axis in the opposite direction of the x-axis and the z-axis parallel to the z-axis of the link coordinate system. With relative translation (5, 3, 0) the matrix is:

$$t_{SL} = \begin{bmatrix} 0 & -1 & 0 & 5 \\ 1 & 0 & 0 & 3 \\ 0 & 0 & 1 & 0 \end{bmatrix}$$

The matrix multiplication of Eq. (3.4) is carried out according to the rules of reduced matrix multiplication from Eq. (2.57).

$$\boldsymbol{R} = \begin{bmatrix} 1 & 0 & 0 \\ 0 & 1 & 0 \\ 0 & 0 & 1 \end{bmatrix} \begin{bmatrix} 0 & -1 & 0 \\ 1 & 0 & 0 \\ 0 & 0 & 1 \end{bmatrix} = \begin{bmatrix} 0 & -1 & 0 \\ 1 & 0 & 0 \\ 0 & 0 & 1 \end{bmatrix}$$

and

$$\boldsymbol{S} = \begin{bmatrix} 1 & 0 & 0 \\ 0 & 1 & 0 \\ 0 & 0 & 1 \end{bmatrix} \cdot \begin{bmatrix} 5 \\ 3 \\ 0 \end{bmatrix} + \begin{bmatrix} 5 \\ 5 \\ 0 \end{bmatrix} = \begin{bmatrix} 10 \\ 8 \\ 0 \end{bmatrix}$$

resulting in

$$t_{SG} = \begin{bmatrix} \boldsymbol{R} & \boldsymbol{S} \end{bmatrix} = \begin{bmatrix} 0 & -1 & 0 & 10 \\ 1 & 0 & 0 & 8 \\ 0 & 0 & 1 & 0 \end{bmatrix}$$

This may easily be verified by inspection, refer to Fig. 3.4.

To transform the coordinates of the super element nodes back to the local coordinate system, matrix t_{SG} must be inverted. This is done according to reduced matrix inversion of Eq. (2.65).

$$\boldsymbol{R}^T = \begin{bmatrix} 0 & 1 & 0 \\ -1 & 0 & 0 \\ 0 & 0 & 1 \end{bmatrix}$$

$$
\begin{aligned}
-\boldsymbol{R}^T\boldsymbol{S} &= \begin{bmatrix} 0 & -1 & 0 \\ 1 & 0 & 0 \\ 0 & 0 & -1 \end{bmatrix} \begin{bmatrix} 10 \\ 8 \\ 0 \end{bmatrix} \\
&= \begin{bmatrix} -8 \\ 10 \\ 0 \end{bmatrix}
\end{aligned}
$$

resulting in

$$
\boldsymbol{t}_{SG}^{-1} = \left[\ \boldsymbol{R}^T - \boldsymbol{R}^T\boldsymbol{S}\ \right] = \begin{bmatrix} 0 & 1 & 0 & -8 \\ -1 & 0 & 0 & 10 \\ 0 & 0 & 1 & 0 \end{bmatrix}
$$

The local coordinates for the two points are now calculated using Eq. (2.59).

$$
\boldsymbol{P}_A = \begin{bmatrix} 0 & 1 & 0 \\ -1 & 0 & 0 \\ 0 & 0 & 1 \end{bmatrix} \begin{bmatrix} 8 \\ 9 \\ 0 \end{bmatrix} + \begin{bmatrix} -8 \\ 10 \\ 0 \end{bmatrix} = \begin{bmatrix} 1 \\ 2 \\ 0 \end{bmatrix}
$$

and

$$
\boldsymbol{P}_B = \begin{bmatrix} 0 & 1 & 0 \\ -1 & 0 & 0 \\ 0 & 0 & 1 \end{bmatrix} \begin{bmatrix} 12 \\ 14 \\ 0 \end{bmatrix} + \begin{bmatrix} -8 \\ 10 \\ 0 \end{bmatrix} = \begin{bmatrix} 6 \\ -2 \\ 0 \end{bmatrix}
$$

The result should be verified by inspection.

The local super element points should now be included as part of the substructure nodes, and kept as external nodes during reduction of the substructure. For simple geometries, as in Example 3.2, the local external points may be set up without any inverse transformation as in the example. However, in the general case with complex space geometries it is difficult to find these points by inspection, especially if we require good accuracy for the following assembling process.

In this section we have shown how to model the position and orientation of joints in a mechanism, and how these points are transformed back to local external nodes for super elements. Other points of interest for the mechanism may be nodes for connecting springs and dampers, nodes for applying external forces, nodes for positioning of additional masses, nodes for applying motion, etc. The modeling of this kind of mechanism components will be covered later, nevertheless, they are mentioned here because substructures must have external nodes corresponding to these points as well. These nodes are handled in much the same way as external nodes for joints both at the mechanism and substructure levels.

3.2 Substructures and Super Elements

3.2.1 Preliminary FE Modeling

For dynamic simulation of flexible mechanisms, each link in the mechanism must be assembled from finite element (FE) models representing one or more substructures/links. In an early stage during a design process, very few details of a product are decided upon, and very coarse finite element models are usually employed in the initial simulation model. Beam elements may often be a good finite element selection for this preliminary modeling of the links, mainly because this will simplify the CAD/FE modeling of the substructures substantially. Also, at this stage a detailed geometric model of the mechanism is usually not available from a CAD system, and the beam element model may be the first iteration of a process for designing mechanical parts. In some cases the first FE model may have very few similarities with the actual geometry of the mechanism links other than the position of the external nodes that later are going to be mechanism joints. This is simply because either the geometry of the links is not designed in any detail yet, or the geometry is too complex for more detailed modeling for this initial simulation. However, in many cases the initial beam model will give a good indication of the geometry of the mechanism, and this beam model is therefore often used for graphic visualization of mechanism motion.

3.2.2 Element Matrices

Two-dimensional beam elements

This finite element will have one node at each end of the element with two translational and one rotational DOFs in each node, see Fig. 3.5. The general finite element equation may now be written:

$$\boldsymbol{S} = \boldsymbol{k}\boldsymbol{v} \tag{3.5}$$

where $\boldsymbol{S}$ is the element force vector (6×1), $\boldsymbol{k}$ is the element stiffness matrix (6×6) and $\boldsymbol{v}$ is the element displacement vector (6×1). For the 2D beam element in expanded form, this equation may be written

$$\begin{bmatrix} S_{x1} \\ S_{y1} \\ S_{rz1} \\ S_{x2} \\ S_{y2} \\ S_{rz2} \end{bmatrix} = \begin{bmatrix} \frac{EA}{l} & 0 & 0 & -\frac{EA}{l} & 0 & 0 \\ 0 & \frac{12EI}{l^3} & \frac{6EI}{l^2} & 0 & -\frac{12EI}{l^3} & \frac{6EI}{l^2} \\ 0 & \frac{6EI}{l^2} & \frac{4EI}{l} & 0 & -\frac{6EI}{l^2} & \frac{2EI}{l} \\ -\frac{EA}{l} & 0 & 0 & \frac{EA}{l} & 0 & 0 \\ 0 & -\frac{12EI}{l^3} & -\frac{6EI}{l^2} & 0 & \frac{12EI}{l^3} & -\frac{6EI}{l^2} \\ 0 & \frac{6EI}{l^2} & \frac{2EI}{l} & 0 & -\frac{6EI}{l^2} & \frac{4EI}{l} \end{bmatrix} \begin{bmatrix} v_{x1} \\ v_{y1} \\ v_{rz1} \\ v_{x2} \\ v_{y2} \\ v_{rz2} \end{bmatrix}$$

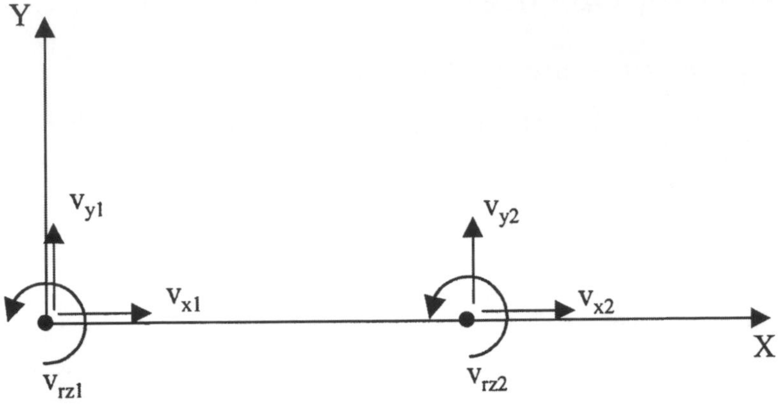

Figure 3.5: Two-dimensional beam element.

where E is the module of elasticity, A is the area of the beam cross section, l the length of the beam, and I is the area moment of inertia.

The derivation of the stiffness matrix $\boldsymbol{k}$ may be found in standard textbooks on the subject. This beam element will also have a corresponding mass matrix of the same dimension (6×6). This mass matrix may be consistent, that is a full matrix derived in a similar way as the stiffness matrix, or lumped, that is the beam mass is divided proportionally between the element nodes. A lumped mass matrix is diagonal.

Three-Dimensional Finite Elements

A general finite element 3D node will have 6 DOFs, 3 translations along the orthogonal axis (x, y, z) and 3 rotations around the same axis. Consequently 3D beam elements with one node in each end will have stiffness and mass matrices with dimensions (12×12). The corresponding element force and displacement vectors, respectively, will have the following components:

$$\boldsymbol{S} = \begin{bmatrix} S_{x1} & S_{y1} & S_{z1} & S_{rx1} & S_{ry1} & S_{rz1} & S_{x2} & S_{y2} & S_{z2} & S_{rx2} & S_{ry2} & S_{rz2} \end{bmatrix}^T \tag{3.6}$$

$$\boldsymbol{v} = \begin{bmatrix} v_{x1} & v_{y1} & v_{z1} & v_{rx1} & v_{ry1} & v_{rz1} & v_{x2} & v_{y2} & v_{z2} & v_{rx2} & v_{ry2} & v_{rz2} \end{bmatrix}^T$$

A triangular and quadrangular shell element will have 3 and 4 element nodes, respectively and corresponding element matrices with dimensions (18×18) and (24×24).

Volume finite elements, however, will only have 3 DOFs in each node, that is the 3 orthogonal translations. A volume element may consequently not transmit torque through only one node.

Joints in a mechanism will transmit torques around the axes where constraints are imposed, and volume element nodes are therefore not suited to model nodes for this kind of joints. In cases where mechanism links are modeled by volume elements, beam or shell elements are usually used to model the node representing the joint. For nodes that only connect volume elements, rotational DOFs are eliminated from the model.

3.2.3 External and Internal Nodes

The finite element model of a substructure will consist of a larger or smaller number of FE nodes, depending on the complexity and detail in the modeling. These nodes will be divided in external and internal nodes, where external nodes will be referred to in the mechanism modeling for joints, springs, dampers, external loads, control input, point of interest etc. These nodes will be retained as so-called super nodes during the reduction of the substructure to a super element. As will be shown in the following internal nodes are eliminated from the FE model.

There should be as few as possible external nodes to reduce the size of the simulation model at system level. To limit model size there may be cases where nodes are split so that translational DOFs are kept external while the rotational DOFs are internal. This type of partly external nodes may be used for axial springs and dampers, external loading, etc.

Example 3.3 : Four-bar mechanism.

Make finite element beam models for the links of a four-bar mechanism, see Fig. 3.6. All links have a width of 0.0254 m, a thickness of 0.0016 m and are made from aluminium.

The length of the input, coupler and output links are 0.0635 m, 0.2794 m and 0.2667 m, respectively. We choose to model each link of this mechanism as one substructure using beam elements. The substructure's finite element meshes are shown in Fig. 3.7.

Here each substructure is modeled from 4 beam elements, and the end nodes (nodes 1 and 5) of each substructure are kept as external nodes.

Suppose we would like to use quadrangular shell elements for substructure 2 in Fig. 3.7, then the finite element model may be

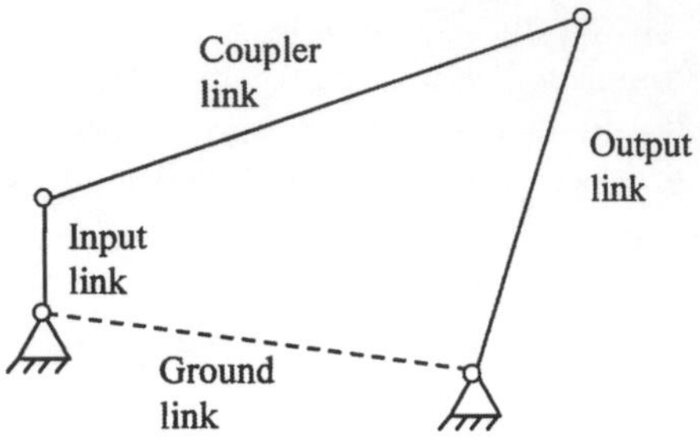

Figure 3.6: Four-bar mechanism.

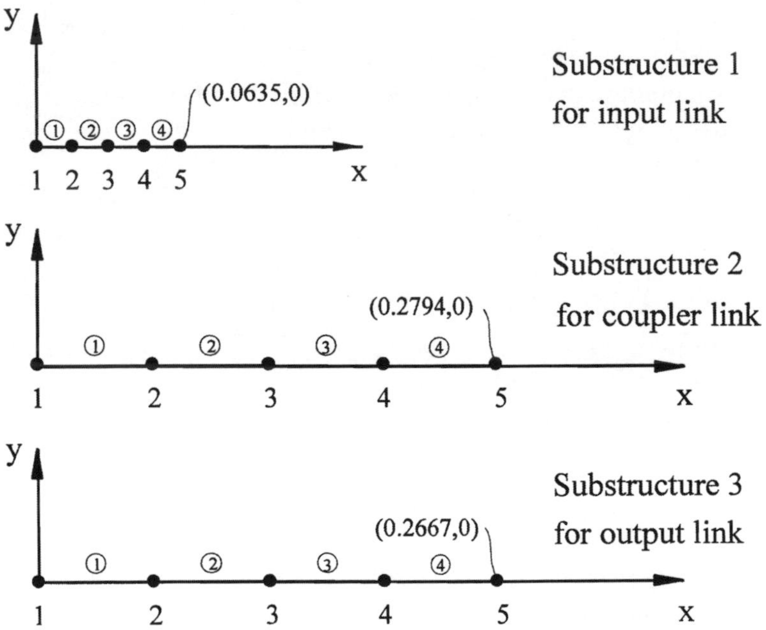

Figure 3.7: Finite element substructure models.

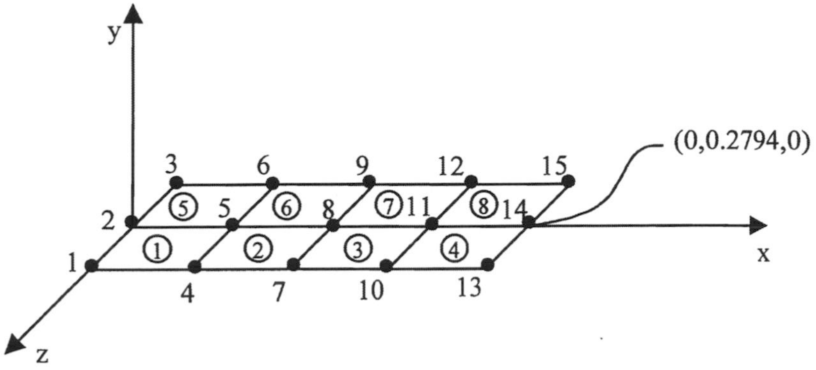

Figure 3.8: Shell finite element mesh for coupler link (substructure 2).

as shown in Fig. 3.8. In this finite element mesh, nodes 2 and 14 are external.

3.2.4 Substructure Matrices and Model Reduction

For each substructure modeled, the status codes for the substructure DOFs are set to 1 for internal DOFs, 2 for external DOFs and 0 for eliminated/constrain DOFs, refer to Section 2.3. Finite element node coordinates are also stored. From this control information the system mass and stiffness matrices are assembled from the corresponding element matrices. The first NDOF1 DOFs of the system matrices are internal substructure DOFs and the next NDOF2 DOFs are external.

In symbolic form the substructure matrices for mass and stiffness may now be divided in submatrices with index i for internal and e for external as follows

$$\begin{bmatrix} \boldsymbol{M}_{ii} & \boldsymbol{M}_{ie} \\ \boldsymbol{M}_{ei} & \boldsymbol{M}_{ee} \end{bmatrix} \quad \text{and} \quad \begin{bmatrix} \boldsymbol{K}_{ii} & \boldsymbol{K}_{ie} \\ \boldsymbol{K}_{ei} & \boldsymbol{K}_{ee} \end{bmatrix}$$

where $\boldsymbol{M}_{xx}$ are mass and $\boldsymbol{K}_{xx}$ are stiffness submatrices. The stiffness relation for the substructure may then be expressed as:

$$\begin{bmatrix} \boldsymbol{K}_{ii} & \boldsymbol{K}_{ie} \\ \boldsymbol{K}_{ei} & \boldsymbol{K}_{ee} \end{bmatrix} \begin{bmatrix} \boldsymbol{v}_i \\ \boldsymbol{v}_e \end{bmatrix} = \begin{bmatrix} \boldsymbol{Q}_i \\ \boldsymbol{Q}_e \end{bmatrix} \tag{3.7}$$

where $\boldsymbol{v}_x$ is the displacement vector and $\boldsymbol{Q}_x$ is the load vector. The first line of Eq. (3.7) may be written:

$$\boldsymbol{v}_i = \boldsymbol{K}_{ii}^{-1}\boldsymbol{Q}_i - \boldsymbol{K}_{ii}^{-1}\boldsymbol{K}_{ie}\boldsymbol{v}_e \tag{3.8}$$

$$= \boldsymbol{v}_i^i + \boldsymbol{v}_i^e \tag{3.9}$$

where $\boldsymbol{v}_i^i$ represents internal displacements with external DOFs fixed and $\boldsymbol{v}_i^e$ represents internal displacements as a function of external displacements.

$$\boldsymbol{v}_i^i = \boldsymbol{K}_{ii}^{-1}\boldsymbol{Q}_i \tag{3.10}$$

and

$$\boldsymbol{v}_i^e = -\boldsymbol{K}_{ii}^{-1}\boldsymbol{K}_{ie}\boldsymbol{v}_e = \boldsymbol{B}\boldsymbol{v}_e \tag{3.11}$$

where

$$\boldsymbol{B} = -\boldsymbol{K}_{ii}^{-1}\boldsymbol{K}_{ie} \tag{3.12}$$

The internal DOFs are going to be eliminated as system DOFs and replaced by a limited number of vibration modes of the substructure called modal DOFs. This reduction of the substructure matrices is called *component mode synthesis* (CMS) transformation, see Langen and Sigbjörnsson (1979). The CMS transformation starts with an eigenvalue analysis of the substructure system matrices with the external DOFs clamped. The equation for free undamped vibration of the substructure internal DOFs may now be written:

$$\boldsymbol{M}_{ii}\ddot{\boldsymbol{v}}_i^i + \boldsymbol{K}_{ii}\boldsymbol{v}_i^i = 0 \tag{3.13}$$

Considering simple harmonic motion the displacement $\boldsymbol{v}_i^i$ may be expressed as:

$$\boldsymbol{v}_i^i = \boldsymbol{\phi}\sin\omega t \tag{3.14}$$

where the eigenvector $\boldsymbol{\phi}$ is defined by the eigenvalue problem.

$$(\boldsymbol{K}_{ii} - \omega^2\boldsymbol{M}_{ii})\boldsymbol{\phi} = \boldsymbol{0} \tag{3.15}$$

Consider now a substructure with n active DOFs of which p are external DOFs. The internal displacements $\boldsymbol{v}_i^i$ may then be expressed as:

$$\boldsymbol{v}_i^i = \sum_{k=1}^{s} \boldsymbol{\phi}_k y_k = \boldsymbol{\Phi}\boldsymbol{y} \tag{3.16}$$

where

$$s < n - p \tag{3.17}$$

and

$$\boldsymbol{\Phi} = \begin{bmatrix} \boldsymbol{\phi}_1 & \boldsymbol{\phi}_2 & \cdots & \boldsymbol{\phi}_S \end{bmatrix} \tag{3.18}$$

is the eigenvector matrix and has dimensions $(n - p) \times s$.

The super element displacements may now be expressed by the external DOFs $\boldsymbol{v}_e$, and by the new modal DOFs $\boldsymbol{y}$:

$$\boldsymbol{v} = \begin{bmatrix} \boldsymbol{v}_e \\ \boldsymbol{v}_i \end{bmatrix} = \begin{bmatrix} \boldsymbol{I} & \boldsymbol{0} \\ \boldsymbol{B} & \boldsymbol{\Phi} \end{bmatrix} \begin{bmatrix} \boldsymbol{v}_e \\ \boldsymbol{y} \end{bmatrix} = \boldsymbol{H} \begin{bmatrix} \boldsymbol{v}_e \\ \boldsymbol{y} \end{bmatrix} \tag{3.19}$$

Usually only a few of the lowest modes of vibration need to be included to get good results, and this may give a substantial reduction in problem size. If all eigenmodes are included, $s = n - p$, then the CMS transformation is exact.

The substructure dynamic equation of motion may be written:

$$\begin{bmatrix} \boldsymbol{M}_{ee} & \boldsymbol{M}_{ei} \\ \boldsymbol{M}_{ie} & \boldsymbol{M}_{ii} \end{bmatrix} \begin{bmatrix} \ddot{\boldsymbol{v}}_e \\ \ddot{\boldsymbol{v}}_i \end{bmatrix} + \begin{bmatrix} \boldsymbol{C}_{ee} & \boldsymbol{C}_{ei} \\ \boldsymbol{C}_{ie} & \boldsymbol{C}_{ii} \end{bmatrix} \begin{bmatrix} \dot{\boldsymbol{v}}_e \\ \dot{\boldsymbol{v}}_i \end{bmatrix} + \begin{bmatrix} \boldsymbol{K}_{ee} & \boldsymbol{K}_{ei} \\ \boldsymbol{K}_{ie} & \boldsymbol{K}_{ii} \end{bmatrix} \begin{bmatrix} \boldsymbol{v}_e \\ \boldsymbol{v}_i \end{bmatrix} = \begin{bmatrix} \boldsymbol{Q}_e \\ \boldsymbol{Q}_i \end{bmatrix} \tag{3.20}$$

where $\boldsymbol{C}_{xx}$ represents damping.

Combining Eq. (3.19) and its first and second time derivatives with Eq. (3.20 and pre-multiplying with $\boldsymbol{H}^T$ gives

$$\begin{bmatrix} \boldsymbol{m}_{11} & \boldsymbol{m}_{12} \\ \boldsymbol{m}_{21} & \boldsymbol{m}_{22} \end{bmatrix} \begin{bmatrix} \ddot{\boldsymbol{v}}_e \\ \ddot{\boldsymbol{y}} \end{bmatrix} + \begin{bmatrix} \boldsymbol{c}_{11} & \boldsymbol{c}_{12} \\ \boldsymbol{c}_{21} & \boldsymbol{c}_{22} \end{bmatrix} \begin{bmatrix} \dot{\boldsymbol{v}}_e \\ \dot{\boldsymbol{y}} \end{bmatrix} + \begin{bmatrix} \boldsymbol{k}_{11} & \boldsymbol{k}_{12} \\ \boldsymbol{k}_{21} & \boldsymbol{k}_{22} \end{bmatrix} \begin{bmatrix} \boldsymbol{v}_e \\ \boldsymbol{y} \end{bmatrix} = \begin{bmatrix} \boldsymbol{q}_1 \\ \boldsymbol{q}_2 \end{bmatrix} \tag{3.21}$$

where

$$\boldsymbol{m}_{11} = \boldsymbol{M}_{ee} + \boldsymbol{B}^T\boldsymbol{M}_{ie} + \boldsymbol{M}_{ei}\boldsymbol{B} + \boldsymbol{B}^T\boldsymbol{M}_{ii}\boldsymbol{B} \tag{3.22}$$

$$\boldsymbol{m}_{12} = \boldsymbol{m}_{21}^T = \boldsymbol{M}_{ei}\boldsymbol{\Phi} + \boldsymbol{B}^T\boldsymbol{M}_{ii}\boldsymbol{\Phi} \tag{3.23}$$

$$\boldsymbol{m}_{22} = \boldsymbol{\Phi}^T\boldsymbol{M}_{ii}\boldsymbol{\Phi} \tag{3.24}$$

$$\boldsymbol{c}_{11} = \boldsymbol{C}_{ee} + \boldsymbol{B}^T\boldsymbol{C}_{ie} + \boldsymbol{C}_{ei}\boldsymbol{B} + \boldsymbol{B}^T\boldsymbol{C}_{ii}\boldsymbol{B} \tag{3.25}$$

$$\boldsymbol{c}_{12} = \boldsymbol{c}_{21}^T = \boldsymbol{C}_{ei}\boldsymbol{\Phi} + \boldsymbol{B}^T\boldsymbol{C}_{ii}\boldsymbol{\Phi} \tag{3.26}$$

$$\boldsymbol{c}_{22} = \boldsymbol{\Phi}^T\boldsymbol{C}_{ii}\boldsymbol{\Phi} \tag{3.27}$$

$$\boldsymbol{k}_{11} = \boldsymbol{K}_{ee} + \boldsymbol{K}_{ie}^T\boldsymbol{B} \tag{3.28}$$

$$\boldsymbol{k}_{12} = \boldsymbol{k}_{21}^T = 0 \tag{3.29}$$

$$\boldsymbol{k}_{22} = \boldsymbol{\Phi}^T\boldsymbol{K}_{ii}\boldsymbol{\Phi} \tag{3.30}$$

$$\boldsymbol{q}_1 = \boldsymbol{Q}_e + \boldsymbol{B}^T\boldsymbol{Q}_i \tag{3.31}$$

$$q_2 = \Phi^T Q_i \tag{3.32}$$

The matrix m_{22} from Eq. (3.24) is diagonal and the eigenmodes may be normalized so that

$$m_{22} = I = \Phi^T M_{ii} \Phi \tag{3.33}$$

that is equal to the unit matrix (Note that m_{11}, m_{12} and m_{21} are not diagonal). The matrix k_{11} from Eq. (3.28) is reduced from the expression

$$k_{11} = K_{ee} + B^T K_{ie} + K_{ei} B + B^T K_{ii} B \tag{3.34}$$

Expanding the last two terms by Eq. (3.12) it may be shown that they are equal but with opposite sign and therefore cancel out in the expression. This is because the symmetry properties of the stiffness matrix give

$$K_{ie}^T = K_{ei} \tag{3.35}$$

and

$$\left(K_{ii}^{-1}\right)^T = K_{ii}^{-1} \tag{3.36}$$

For the same reasons the matrices $k_{12} = k_{21}^T$ are reduced from the equation

$$\begin{aligned} k_{12} &= k_{21}^T = K_{ei}\Phi + B^T K_{ii}\Phi \\ &= K_{ei}\Phi + (-K_{ii}^{-1} K_{ie})^T K_{ii}\Phi \\ &= K_{ei}\Phi - K_{ie}^T \left(K_{ii}^{-1}\right)^T K_{ii}\Phi = 0 \end{aligned} \tag{3.37}$$

It may be shown that the stiffness matrix k_{22} is diagonal and of the form

$$k_{22} = diag\left[\begin{array}{cccc} \omega_1^2 & \omega_2^2 & \cdots & \omega_{n-p}^2 \end{array}\right] \tag{3.38}$$

where ω_1^2, ω_2^2, ... ω_{n-p}^2 are the eigenvalues corresponding to the eigenmodes of eigenvector matrix Φ.

The substructure matrices that are reduced by CMS transformation, as shown above, are the super element matrices in a standardized form. These matrices will later form the basis for the mechanism simulation formulation.

3.3 Updated Super Elements

3.3.1 Corotated Coordinate Systems

During mechanism simulation the super elements generally have very large rotations, and the super element matrices defined in Section 3.2 must be transformed to the actual super element direction for new positions of the mechanism during simulation. The super element matrices are all referred to the corresponding coordinate system where the finite element mesh was modeled at substructure level. In order to calculate the transformations for the super elements for each new position, the direction for the corresponding super element reference coordinate system must be calculated. This system is called the *Super Element Corotated Coordinate System* (SECCS). The only way to calculate the SECCS for new positions during simulations is to refer it to the positions and orientations of super nodes at the actual super element. For each new position during a simulation the super nodes have of course a position, but also an orientation of a coordinate system that is connected rigidly to the node. The super node position and orientation are primary variables of the simulation, and are therefore available at any time during simulation.

Each SECCS may now be connected to a set of super nodes on the actual super element in the following way:

1. A super node on the actual super element is selected to represent the position of the origin of the SECCS. If the super node selected has an offset from the origin of the SECCS, the offset is specified in Cartesian coordinates relative the super node coordinate system. From this specification, a point may be calculated in global coordinates that gives the origin of the SECCS.

2. A second super node on the super element is selected to represent a point on the positive x-axis of the SECCS. If the super node selected has an offset from the x-axis of the SECCS, the offset is specified in Cartesian coordinates relative to the super node coordinate system. From this specification a point may be calculated in global coordinates that gives a point on the SECCS' x-axis.

3. A third and last super node on the super element is selected to represent a point in the xy-plane in the positive y-direction of the SECCS, for instance on the positive y-axis. If the super node selected has an offset from the xy-plane of the SECCS, the offset is specified in Cartesian coordinates relative to the super node coordinate system. From this specification a point may be calculated in global coordinates that gives a point in the xy-plane in the positive y-direction of the SECCS.

The same super node may be referred to more than once. However, at least two different super nodes should be referred to for each super element. The transformation matrix for the SECCS relative to the global coordinate system may now easily be calculated from the three reference points calculated above using vector product operations.

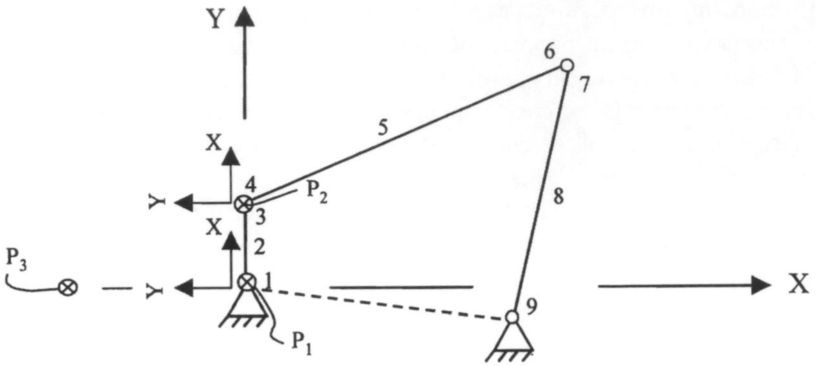

Figure 3.9: Super element corotated coordinate system.

Example 3.4 : Updated super elements.

Specify and calculate the SECCS for the input link of Example 3.3, see also Fig. 3.9

Super nodes 1, 3 and 1 are referred to in that order to specify the three reference points for the SECCS of the input link. The transformation matrix for super node 1, relative the global coordinate system, is:

$$\boldsymbol{t}_{SN1} = \begin{bmatrix} 0 & -1 & 0 & 0 \\ 1 & 0 & 0 & 0 \\ 0 & 0 & 1 & 0 \end{bmatrix}$$

and for super node 3:

$$\boldsymbol{t}_{SN3} = \begin{bmatrix} 0 & -1 & 0 & 0 \\ 1 & 0 & 0 & 0.0635 \\ 0 & 0 & 1 & 0 \end{bmatrix}$$

Reference point 1 is referred to super node 1 with the Cartesian offset relative $\boldsymbol{t}_{SN1}$ of $\boldsymbol{o}_1^T = [0, 0, 0]$

Reference point 2 is referred to super node 3 with the Cartesian offset relative $\boldsymbol{t}_{SN3}$ of $\boldsymbol{o}_2^T = [0, 0, 0]$

Reference point 3 is referred to super node 1 with the Cartesian offset relative $\boldsymbol{t}_{SN1}$ of $\boldsymbol{o}_3^T = [0, 1, 0]$

The three global points $\boldsymbol{P}_1$, $\boldsymbol{P}_2$ and $\boldsymbol{P}_3$ may now be calculated, see the modified transformations in Section 2.2.

$$\boldsymbol{P}_1 = \boldsymbol{t}_{SN1} \cdot \boldsymbol{o}_1 = \begin{bmatrix} 0 \\ 0 \\ 0 \end{bmatrix}$$

$$\boldsymbol{P}_2 = \boldsymbol{t}_{SN3} \cdot \boldsymbol{o}_2 = \begin{bmatrix} 0 \\ 0.0635 \\ 0 \end{bmatrix}$$

$$\boldsymbol{P}_3 = \boldsymbol{t}_{SN1} \cdot \boldsymbol{o}_3 = \begin{bmatrix} -1 \\ 0 \\ 0 \end{bmatrix}$$

The reader should now be able to verify that the transformation matrix for the SECCS in this position is:

$$\boldsymbol{t}_{SE1} = \begin{bmatrix} 0 & -1 & 0 & 0 \\ 1 & 0 & 0 & 0 \\ 0 & 0 & 1 & 0 \end{bmatrix}$$

3.3.2 Deformations and Stiffness Forces

During mechanism simulation, the super nodes will move to new positions due to elastic and rigid body motion of the super elements. To calculate the deformations of one super element we start by calculating the SECCS transformation matrix and its inverse, see Eq. (2.65).

The transformation matrices of all super nodes of the actual super element are then transformed by the inverse super element transformation calculated above, see Fig. 3.10.

The rigid body motion is then eliminated by the inverse super element transformation of the super nodes, and the deviation from the initial undeformed super node positions and orientations may now be used to calculate the elastic deformations of the super element. We are assuming small elastic deformations and rotations of the super nodes, and following inverse super element transformation of a super node transformation, the super node transformation will be in the form (see also Eq. (2.69)):

$$\boldsymbol{t}'_{SNj} = \boldsymbol{t}_{SEi}^{-1} \cdot \boldsymbol{t}_{SNj} = \begin{bmatrix} 1 & -\Delta\gamma & \Delta\beta & x \\ \Delta\gamma & 1 & -\Delta\alpha & y \\ -\Delta\beta & \Delta\alpha & 1 & z \end{bmatrix} \tag{3.39}$$

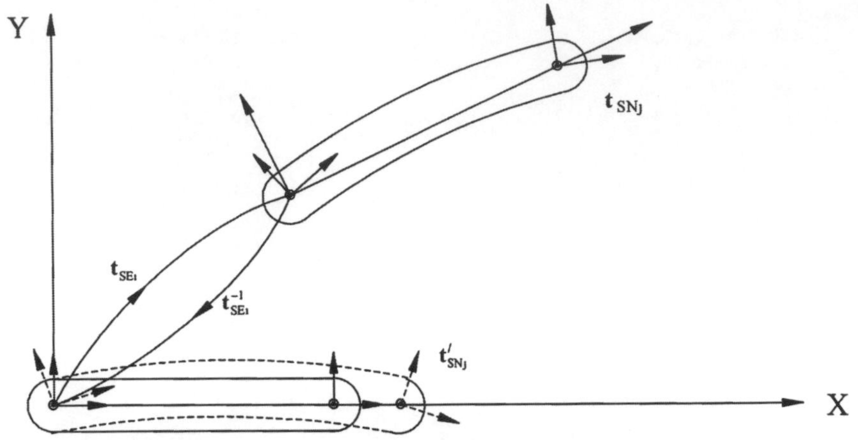

Figure 3.10: Deformations of super elements.

The super node deformation vector for node j may be calculated from

$$\boldsymbol{v}_j = \begin{bmatrix} v_{xj} \\ v_{yj} \\ v_{zj} \\ v_{rxj} \\ v_{ryj} \\ v_{rzj} \end{bmatrix} = \begin{bmatrix} x - x_o \\ y - y_o \\ z - z_o \\ \Delta\alpha \\ \Delta\beta \\ \Delta\gamma \end{bmatrix} \tag{3.40}$$

where x_o, y_o and z_o are the undeformed x-, y- and z-position of the super node.

Repeating the process of Eqs. (3.39) and (3.40) for all super nodes of a super element, and noticing that the modal coordinates from the internal super element modes of vibration are explicit variables in the simulation, the super element deformation vector may be assembled, see Eq. (3.19).

$$\begin{bmatrix} \boldsymbol{v}_e \\ \boldsymbol{y} \end{bmatrix} = \begin{bmatrix} \boldsymbol{v}_1 \\ \boldsymbol{v}_2 \\ \cdot \\ \cdot \\ \boldsymbol{v}_l \\ y_1 \\ y_2 \\ \cdot \\ \cdot \\ y_s \end{bmatrix} \tag{3.41}$$

In this general case the super element has l super node deformations calculated from Eq. (3.40) and s modal coordinates.

The internal super element stiffness forces $\boldsymbol{S}$ are calculated from (see Eqs. (3.5) and (3.21)):

$$\boldsymbol{S} = \begin{bmatrix} \boldsymbol{S}_e \\ \boldsymbol{S}_{mod} \end{bmatrix} = \begin{bmatrix} \boldsymbol{k}_{11} & \boldsymbol{k}_{12} \\ \boldsymbol{k}_{21} & \boldsymbol{k}_{22} \end{bmatrix} \begin{bmatrix} \boldsymbol{v}_e \\ \boldsymbol{y} \end{bmatrix} \tag{3.42}$$

where $\boldsymbol{S}_{mod}$ are forces in modal coordinates.

3.3.3 Assembling into System Matrices

Super node DOFs have global directions as default, but may be specified to have updated local direction. Updated local directions are referenced to the coordinate system attached to the actual super node or to another specified super node. Prior to adding the super element matrices into the system matrix the different nodes will be transformed to the actual global or updated local directions.

Let us suppose that the rotational part of the inverse super element transformation matrix $\boldsymbol{t}_{SEi}^{-1}$ for super element no. i is called $\boldsymbol{R}_{SEi}^T$ and the rotational part of the transformation matrix $\boldsymbol{t}_{SNj}$ for super node j is called $\boldsymbol{R}_{SNj}$. The transformation for super element node k before adding into the system matrices is called $\boldsymbol{R}_k$ and is equal to

$$\boldsymbol{R}_k = \boldsymbol{R}_{SEi}^T \tag{3.43}$$

if the global direction is specified for the node, or

$$\boldsymbol{R}_k = \boldsymbol{R}_{SEi}^T \boldsymbol{R}_{SNj} \tag{3.44}$$

if the local updated direction is specified. Super node j may be the same as node k. The transformation matrix $\boldsymbol{R}_k$ of dimension 3×3 is valid for super element node k both for translation and rotation transformations. For a super element with l external nodes and s modal DOFs, the overall super element transformation matrix may be written:

$$\boldsymbol{T}_{SEi} = \begin{bmatrix} \boldsymbol{R}_1 & & & & & & & & \\ & \boldsymbol{R}_1 & & & & & 0 & & \\ & & \boldsymbol{R}_2 & & & & & & \\ & & & \boldsymbol{R}_2 & & & & & \\ & & & & \ddots & & & & \\ & & & & & \boldsymbol{R}_l & & & \\ & 0 & & & & & \boldsymbol{R}_l & & \\ & & & & & & & \boldsymbol{I}_{mod} \end{bmatrix} \tag{3.45}$$

where the order of $\boldsymbol{I}_{mod}$ is equal to s, that is the number of modal coordinates for the super element.

The mass and stiffness matrices of the super element are denoted $\boldsymbol{m}$ and $\boldsymbol{k}$, respectively, see Eq. (3.21). The super element matrices are transformed to the actual directions for the DOFs at system level as follows

$$\bar{\boldsymbol{m}} = \boldsymbol{T}_{SEi}^T \boldsymbol{m} \boldsymbol{T}_{SEi} \tag{3.46}$$

$$\bar{\boldsymbol{k}} = \boldsymbol{T}_{SEi}^T \boldsymbol{k} \boldsymbol{T}_{SEi} \tag{3.47}$$

where $\bar{\boldsymbol{m}}$ and $\bar{\boldsymbol{k}}$ are the transformed mass and stiffness super element matrices, respectively.

The local reduced super element mass and stiffness matrices may be stored in the computer, and at each new position of the mechanism, these matrices are given the geometric transformation from Eqs. (3.46) and (3.47), respectively. The transformed super element matrices may then be added into the system matrix for mass

$$\boldsymbol{M} = \sum_i \boldsymbol{a}_i^T \bar{\boldsymbol{m}}_i \boldsymbol{a}_i \tag{3.48}$$

and for stiffness

$$\boldsymbol{K} = \sum_i \boldsymbol{a}_i^T \bar{\boldsymbol{k}}_i \boldsymbol{a}_i \tag{3.49}$$

where $\boldsymbol{a}_i$ are incidence matrices that represent the super element topology at system level. The index i runs over all super elements. In an actual computer program the summation is done more efficiently and no matrix multiplication is carried out.

3.3.4 Corotational Geometric Stiffness

The mechanism simulated consists of links and each link consists of one or several elements. The link is regarded as a super element with a variable number of external nodes. The links are connected together through their external nodes to form a mechanism. The mechanism is simulated nonlinearly i.e. the geometry is updated throughout the analysis.

By adding rotational geometric stiffness (pre-stress stiffness) to the material stiffness the stiffness matrix K will improve. If the structure is in a stressed state, we will get a more correct stiffness matrix (closer to the tangent). The convergence rate of the Newton-Raphson iteration algorithm will improve for most cases. This will also give more correct eigenfrequencies for structures in a stressed state.

A complete implementation of all stiffness terms from the corotated formulation could be carried out but we probably would not gain much profit compared the extra CPU time needed. The material stiffness term and the rotational geometric stiffness term are the most important stiffness terms in the tangent stiffness matrix for the corotated formulation when the deformations are small.

The rotational geometric stiffness term is very sensitive to the positioning of the reference points that define the super element's local coordinate system. An automatic selection and optimization algorithm has been developed and implemented.

The tangent stiffness originally only consisted of the material stiffness term, however, there are situations where the rotational geometric stiffness term should be included. For the test examples this led to an increased convergence rate and more correct eigenfrequencies. Generally, these effects will be present but there are examples where the convergence rate is higher with rotational geometric stiffness included. For details refer to Appendix A

3.4 Master-Slave Constrained Joints

A joint is a way of specifying a constrained relative motion between two bodies or links in a mechanism. In the finite element modeling of mechanism joints, super element nodes are used to specify the joint constraints. A joint between two moving links, A and B, will in general be defined by one node on link A called the slave node, and one or more nodes on link B called the master nodes. The constraints are then modeled by making DOFs of the slave node dependent on the corresponding DOFs of the master node(s).

The dependent (slave) DOFs representing constraints of a joint are expressed as a linear combination of a number ($\geq$1) of free (master) DOFs, refer to Eq. (2.83):

$$r_i = r_{slave} = c_0 + \sum_{j=1}^{P} c_j r_{master\,j} \tag{3.50}$$

where c_0, c_1, c_2 ... c_p are coefficients that may be constants or variables depending on the joint type. For Eq. 3.50 to be valid, all the involved DOFs must be transformed to the same local updated direction, see Eq. 3.44.

3.4.1 The Revolute Joint

A revolute joint is modeled from two coincident super nodes located on different links in a mechanism. The coordinate systems attached to these two super nodes must have a common z-axis coincident with the current joint

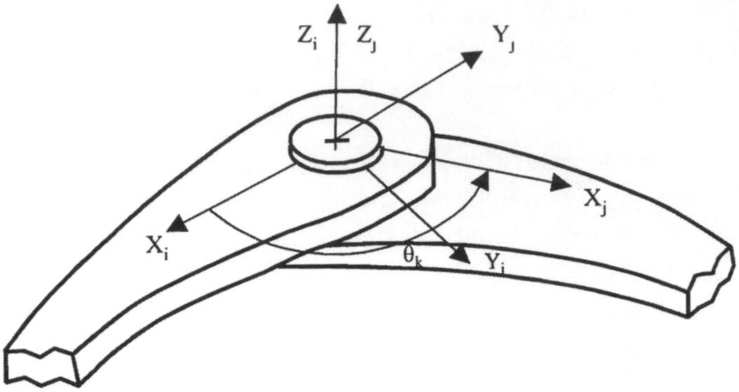

Figure 3.11: Coordinate axis and variable of revolute joint.

axis of revolution. The coordinate system i is attached to the master node while j is attached to the slave or dependent node, see Fig. 3.11.

The joint has one degree of freedom represented by the variable θ_k in Fig. 3.11. The DOFs of these two nodes are transformed to the same updated, local direction, for instance the current direction of coordinate system i, during equation solution. These two nodes will produce 7 active DOFs in the system equation, all six DOFs of the master node plus the rotation about z for the slave node. The joint variable θ_k is derived from the relative rotations about the z-axis, see Eq. (2.82)

$$\theta_k = \theta(\boldsymbol{t}_i^{-1}\boldsymbol{t}_j) \tag{3.51}$$

The joint produces five constraint equations with $c_0 = 0$, $P = 1$ and $c_1 = 1$, see Eq. (3.50).

$$\begin{aligned} x_j &= x_i \\ y_j &= y_i \\ z_j &= z_i \\ x_{rj} &= x_{ri} \\ y_{rj} &= y_{ri} \end{aligned} \tag{3.52}$$

where x, y and z are translations along the corresponding axis, while x_r and y_r are rotations around the x- and y-axis, respectively. This implies that the left-hand side variables of Eq. (3.52) are dependent variables that are eliminated during the equation solution for time integration.

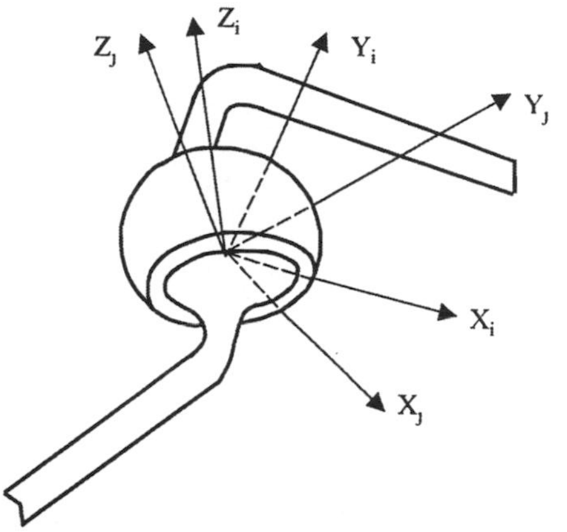

Figure 3.12: Coordinate system i and j of a ball joint.

3.4.2 The Ball Joint

A ball joint is modeled from two coincident super nodes located on different links in a mechanism. The ball joint has no joint axis, and hence, the only requirements are that the origin of the attached coordinate systems of the super nodes must coincide. The coordinate system i is attached to the master node while j is attached to the slave or dependent node, see Fig. 3.12.

The three rotational variables between the two coordinate systems attached to each link of the ball joint may be defined in different ways, for instance as Euler Angles, see Eq. (2.82). The DOFs of these two nodes are transformed to the same global or updated local direction, for instance the current direction of coordinate system i, during equation solution. These two nodes, the master (i) and the slave (j) node, will produce 9 active DOFs in the system equation, all six DOFs of the master node plus the three rotations of the slave node. The joint produces three constraint equations with $c_0 = 0$, $P = 1$ and $c_1 = 1$, see Eq. (3.50).

$$\begin{aligned} x_j &= x_i \\ y_j &= y_i \\ z_j &= z_i \end{aligned} \tag{3.53}$$

where x, y and z are translations along the corresponding axis. This implies that the left-hand side variables of Eq. (3.53) are dependent variables that are eliminated during equation solution for time integration.

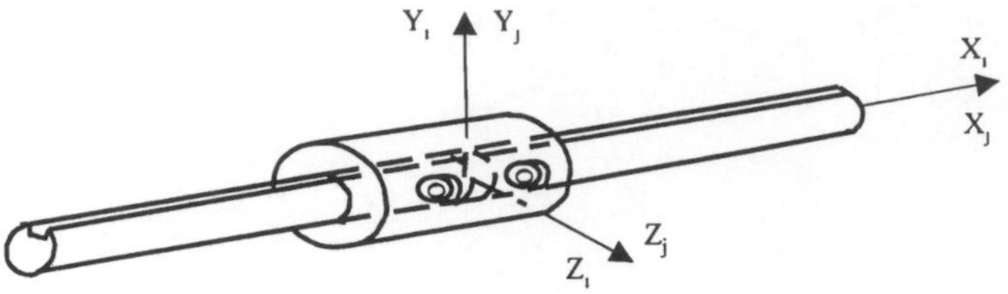

Figure 3.13: Coordinate axes in a rigid joint.

3.4.3 The Rigid Joint

A rigid joint is modeled from two coincident super nodes located on different links in a mechanism. The coordinate systems attached to these two super nodes must also be coincident, but with no limitation of direction in space as long as they are common. The coordinate system i is attached to the master node while j is attached to the slave or dependent node, see Fig. 3.13. The DOFs of these two nodes are transformed to global or to the same updated local direction, for instance the current direction of coordinate system i, during equation solution. The two nodes together, the master (i) and the slave (j) node will produce 6 active DOFs in the system equation, all from the master node. The joint has no joint variable. It produces six constraint equations with $c_0 = 0$, $P = 1$ and $c_1 = 1$, see Eq. (3.50):

$$\begin{aligned} x_j &= x_i \\ y_j &= y_i \\ z_j &= z_i \\ x_{rj} &= x_{ri} \\ y_{rj} &= y_{ri} \\ z_{rj} &= z_{ri} \end{aligned} \tag{3.54}$$

where x, y and z are translations along the corresponding axis, while x_r, y_r and z_r are rotations around the corresponding axis. This implies that the left-hand side variables of Eq. (3.54) are dependent variables that are eliminated during equation solution for time integration.

The rigid joint may be used for accessing internal forces through the joint at system level during simulation.

3.4.4 The Free Joint

The free joint is modeled from two super nodes located on different links in a mechanism. A coordinate system i is attached to the master node while j

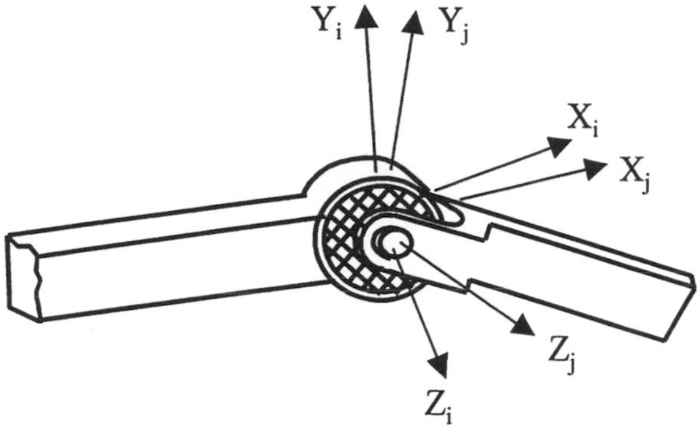

Figure 3.14: Coordinate system i and j for a free joint.

is attached to the slave node, see Fig. 3.14.

That is, these two nodes, the master (i) and the slave (j), will produce 12 active DOFs in the system equation, all six from both nodes. The joint has 6 degrees of freedom with 6 corresponding joint variables, translation in x-, y- and z-directions and rotation around the same axis, all referred to the coordinate system attached to the master node. For definition of rotation variables, refer to Eq. (2.82). Though it may be argued that the term "slave node" is not relevant for this joint because no constraint equations are introduced, the term is used to have a systematic approach to joint modeling. However, it will be shown later that constraints may be introduced by springs for this joint type.

3.4.5 The Prismatic Joint

The prismatic joint is modeled from three or more super nodes where one node, the slave node, is located on the first link while two or more nodes, the master nodes, are located on the second link. The master nodes define a curve for where the sliding is to occur. The coordinate systems attached to the slave and master nodes must have the same direction in space, see Fig. 3.15.

The DOFs of all the involved nodes in a prismatic joint (Fig. 3.15) are transformed to the same updated local direction, for instance the current direction of coordinate system i, during equation solution. The master nodes, node 1, ..., k, ..., (number of masters), will each produce 6 DOFs in the system equation, while the slave node will produce 3 DOFs, the translation in z-direction and the rotations about the x- and y-axes. The joint variable s_k is

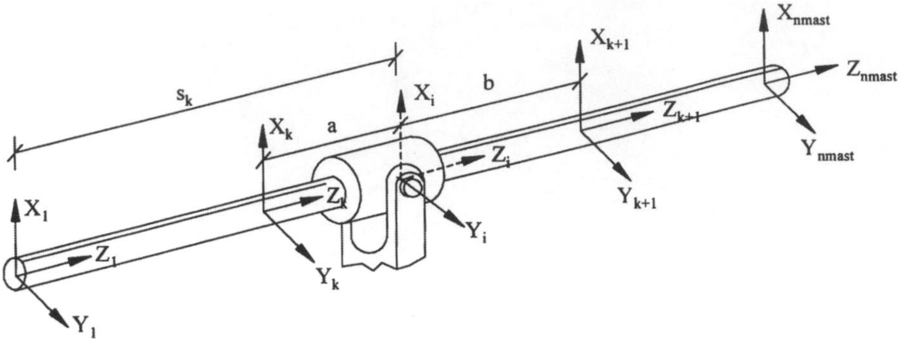

Figure 3.15: Coordinate system attached to the slave node (i) and to the master nodes for the prismatic joint.

derived from the relative translation along the z-axis.

$$s_k = s(\boldsymbol{t}_1^{-1}\boldsymbol{t}_i) \tag{3.55}$$

The joint produces three constraint equations with $P =$ (number of masters) and with $c_0 = c_1 = c_2 = \ldots = c_{\text{number of masters}} = 0$ except $c_k = 1 - \alpha$ and $c_{k+1} = \alpha$, see Eq. (3.50) and Fig. 3.15($\alpha = a/(a+b)$):

$$\begin{aligned} x_i &= c_k x_k + c_{k+1} x_{k+1} \\ y_i &= c_k y_k + c_{k+1} y_{k+1} \\ z_{ri} &= c_k z_{rk} + c_{k+1} z_{r(k+1)} \end{aligned} \tag{3.56}$$

where x and y are translations along the corresponding axes, while z_r is rotation around the z-axis. This implies that the left-hand side variables of Eq. (3.56) are dependent variables that are eliminated during equation solution for time integration. To model telescopic motion, two parallel prismatic joints of this type are used between the same two links of the mechanism.

3.4.6 Other Joint Types

The cylindrical joint may be modeled in a very similar way as the prismatic joint, however, the rotation around the joint axis will also be a joint variable. Universal joints are modeled in detail using two or more revolute joints.

3.5 Master-Slave Constrained Transmissions

The transmission elements described in the following are based on the master-slave technique presented in Section 2.3. See also joint modeling in the previous section.

3.5.1 Gear Joint

The gear joint is based on two revolute joints, one for the gear input shaft and one for the gear output shaft. For these two joints the master nodes must be positioned on the same link - the gear housing. The degree of freedom statuses for the joint of the gear output shaft are typically for the slave and the master node, respectively:

	x_t	y_t	z_t	x_r	y_r	z_r
S_O	0	0	0	0	0	1
M_O	1	1	1	1	1	1

where

0: represent dependent DOF

1: represent independent DOF

t: the subscript refers to translational DOF

r: the subscript refers to rotational DOF

S: slave node

M: master node

O: the subscript refers to output shaft

The degree of freedom statuses for the joint of the gear input shaft are similarly for the slave and the master node, respectively:

	x_t	y_t	z_t	x_r	y_r	z_r
S_I	0	0	0	0	0	1
M_I	1	1	1	1	1	1

where

I: the subscript refers to input shaft

The rotation of the gear output shaft is represented by the rotational DOF around the z-axis for the node S_O, see above. This rotation will now be represented as a linear combination of the rotation about the z-axis of the input shaft and the rotation about the z-axis of the master nodes on the gear housing (M_O and M_I).

Setting the gear ratio to N, these linear dependencies may be written:

$$(S_O - M_O) * N = (S_I - M_I) \tag{3.57}$$

that may be rewritten:

$$S_O = M_O + S_I/N - M_I/N \tag{3.58}$$

During simulation the correct angle of the gear output shaft may be calculated from

$$(v_{OC} - v_{O0}) * N = (v_I - v_{I0}) \tag{3.59}$$

that may be written

$$v_{OC} = v_{O0} + v_I/N - v_{I0}/N \tag{3.60}$$

where

v_I: gear input shaft angle

v_{I0}: initial gear input shaft angle

v_{OC}: calculated gear output shaft angle

v_{O0}: initial gear output shaft angle

N: gear ratio

We may now calculate the correction for the output gear angle from:

$$\Delta v = v_{OC} - v_O \tag{3.61}$$

where

v_O: current gear output shaft angle

Δv: correction for the current gear output shaft angle

The degree of freedom statuses for the involved nodes shown above will be kept except for the slave node of the output shaft where the rotation about the z-axis will be dependent, that is for the S_O node the DOF statuses will be

	x_t	y_t	z_t	x_r	y_r	z_r
S_O	0	0	0	0	0	0

and the coefficients in the table of constraint coefficients will be

$$TCC = [0, 1/N, 1, -1/N]$$

in the scquence S_O, S_I, M_O and M_I.

3.5.2 Rack and Pinion Joint

The rack and pinion joint is based on one prismatic joint and one revolute joint, the prismatic joint for the linear output movement and the revolute joint for the input shaft. The degree of freedom statuses for the output prismatic joint of the rack and pinion is typically for the slave and the first master node, respectively:

	x_t	y_t	z_t	x_r	y_r	z_r
S_O	0	0	1	1	1	0
M_{O1}	1	1	1	1	1	1

The degree of freedom statuses for the rack and pinion input shaft are similarly for the slave and the master node, respectively:

	x_t	y_t	z_t	x_r	y_r	z_r
S_I	0	0	0	0	0	1
M_I	1	1	1	1	1	1

The linear translation of the rack and pinion is represented by the translational DOF along the z-axis for the node S_O, see above. This translation will then be represented as a linear combination of the rotation about the z-axis of the slave node of the input shaft and the translation along and rotation about, respectively, the z-axis of the master nodes referred to (M_{O1} and M_I).

Setting the gear ratio to N, these linear dependencies may be written:

$$(S_O - M_{O1}) * N = (S_I - M_I) \tag{3.62}$$

and may be rewritten:

$$S_O = M_{O1} + S_I/N - M_I/N \tag{3.63}$$

During simulation the correct position of the rack and pinion may be calculated from

$$(s_{OC} - s_{O0}) * N = (v_I - v_{I0}) \tag{3.64}$$

that may be written:

$$s_{OC} = s_{O0} + v_I/N - v_{I0}/N \tag{3.65}$$

where

v_I: transmission input shaft angle

v_{I0}: initial transmission input shaft angle

s_{OC}: calculated rack and pinion position

s_{O0}: initial rack and pinion position

N: gear ratio

We may now calculate the correction for the rack and pinion position from:

$$\Delta s = s_{OC} - s_O \tag{3.66}$$

where

s_O: current rack and pinion position

Δs: correction for the current rack and pinion position

The degree of freedom statuses for the involved nodes shown above will be kept except for the slave node of the prismatic joint where the translation along the z-axis will be dependent, that is for the S_O node the DOF statuses will be:

	x_t	y_t	z_t	x_r	y_r	z_r
S_O	0	0	0	1	1	0

and the coefficients in the table of constraint coefficients will be:

$$TCC = [0, 1/N, 1, -1/N]$$

in the sequence S_O, S_I, M_{O1} and M_I.

3.5.3 Screw Joint

The screw joint is based on a cylindrical joint. The degree of freedom statuses for the cylindrical joint of the screw transmission is typically for the slave and the first master nodes, respectively:

	x_t	y_t	z_t	x_r	y_r	z_r
S	0	0	1	1	1	1
M_1	1	1	1	1	1	1

The linear translation of the rack and pinion is represented by the translational DOF along the z-axis for the node S, see above. This translation will now be represented as a linear combination of the rotation about the z-axis of node S and the translation along and rotation about, respectively, the z-axis of the master nodes referred to (M_1).

Setting the gear ratio to N, these linear dependencies may be written:

$$(S - M_1) * N = (S_r - M_{r1}) \tag{3.67}$$

where

S: translation along z-axis of node S

M_1: translation along z-axis of node M1

S_r: rotation about z-axis of node S

M_{r1}: rotation about z-axis of node M1

this may be rewritten:

$$S = M_1 + S_r/N - M_{r1}/N \tag{3.68}$$

During simulation the correct position of the screw may be calculated from

$$(s_C - s_0) * N = (v_r - v_{r0}) \tag{3.69}$$

that may be written:

$$s_C = s_0 + v_r/N - v_{r0}/N \tag{3.70}$$

where

v_r: screw angle

v_{r0}: initial screw angle

s_C: calculated screw position

s_0: initial screw position

N: gear ratio

We may now calculate the correction for the screw position from:

$$\Delta s = s_C - s \tag{3.71}$$

where

s: current screw position

Δs: correction for the current screw position

The degree of freedom statuses for the involved nodes shown above will be kept except for the slave node where the translation along the z-axis will be dependent, that is for the node S the DOF statuses will be:

	x_t	y_t	z_t	x_r	y_r	z_r
S	0	0	0	1	1	1

and the coefficients in the table of constraint coefficients will be:

$$TCC = [0, 1/N, 1, -1/N]$$

in the sequence S, S_r, M_1 and M_{r1}.

3.6 Flexible Connections

3.6.1 Springs and Dampers

Spring Elements

A spring may be linearly defined by a spring constant or it may be nonlinear where the spring stiffness is defined by a reference to an explicit function definition. For nonlinear springs the stiffness is a function of the translational or angular spring deflection. Effects like endstops, backlash in gears and other nonlinear effects may easily be modeled in this way. An axial spring is specified between two super nodes in the mechanism model while joint springs are specified on joint degrees of freedom.

The axial spring element is handled in a similar way as a beam or truss element, that is the stiffness is transformed to the actual direction of the two connecting nodes and added into the system matrix.

For joint springs, however, the stiffness is added directly to the diagonal of the system matrix on the actual two DOFs involved and to their coupling elements.

A spring may be used as a passive element with fixed stress free length or angle, however, it may also be used as an active drive element in the mechanism where the stress free length varies as a function of time in accordance with an explicitly defined function.

By specifying springs on joint DOFs, as mentioned earlier, special constraint effects may be imposed, for instance the free joint may be used in this way to model nonlinear bearings or rubber bushings.

Damping Elements

A damper may be linear defined by a damping constant or it may be nonlinear where the damping coefficient is defined by a reference to an explicit function definition. For nonlinear dampers the damping coefficient is a function of the translational or angular damper velocity. A wide range of energy dissipation effects may be modeled by damping elements. An axial damper is specified between two super nodes in the mechanism model while joint dampers are specified on joint degrees of freedom.

As for spring elements, the axial damper element is handled in a similar way as a beam or truss element, that is the damping coefficient is transformed to the actual direction of the two connecting nodes and added into the system matrix.

For joint dampers, however, the damping coefficient is added directly to the diagonal of the system matrix on the actual two DOFs involved and their coupling elements.

3.6.2 Spring Constrained Joints

All joint variables for master-slave based joints such as revolute joints, ball joints, prismatic joints, etc. may be constrained by springs. The so-called free joint has no degree of freedom constraint and is usually the basis for modeling spring constrained joints. The reason for making joints spring constrained can be to model the flexibility of an actual bearing or model nonlinear effects like clearings, rubber bushings, etc. To be able to switch easily between master-slave constrained joints and spring constrained joints the position and orientation of triads for spring constrained joints should be the same as for the corresponding master-slave constrained joint.

The *flexible revolute joint* is usually modeled from a free joint with relative constraining springs on the x, y and z translational variables and on the x and y rotational joint variables. The rotation about the z-axis will be the joint variable. The stiffness of the springs on the different joint variables should comply with the actual translational and rotational stiffness of the bearing. Nonlinear stiffness like in rubber bushings or in bearings with clearing should be modeled by nonlinear springs.

The *flexible ball joint* will also usually be based on a free joint and with constraining springs for the x, y and z translational variables. The stiffness for the springs will be set according to joint stiffness and possible nonlinearities as for the revolute joint above.

The *flexible prismatic joint* will usually be modeled from one or more free joints. The z-axis of the joint triads should be along the direction of the joint variable, the joint axis. If the flexible prismatic joint is based on only one free joint all joint variables should have constraining springs except the variable along the joint axis, that is the z direction of the joint.

If two or more free joints are used to model a flexible prismatic joint the orientation of the corresponding triads shoud be the same and with the z-direction along the joint axis. For the free joints constituting the flexible prismatic joint the joint variables in x and y directions and rotation about z should have constraining springs. The rotation variables about the x and y directions for the actual free joints will usually have no constraining springs, see the master-slave based prismatic joint. Translational clearing in the joint could be modeled by nonlinear constraining springs for x and/or y translation and rotational clearing about the z axis is modeled by nonlinear constraining springs for the z rotation.

The *flexible cylindric joint* is modeled in a very simular way as the flexible prismatic joint described above. However, there is no constraining spring for the z rotation. This joint will have two joint variables, the translation along and the rotation about the joint axis that is the z-axis of the triads.

Besides the options for modeling constraints mentioned above, users have almost unlimited options for modeling different constrained motion, however, they should be aware of what effects could result from the fact that rotation in space does not commute, see Section 2.2.2.

3.6.3 Spring Constrained Transmissions

Mechanical transmissions are often a major source of mechanical deflections or compliance. Poor mechanical stiffness does not only cause static deflections but also limits dynamic response. Poor stiffness at for example gearing may also cause undesirable vibrations and is a critical issue in high performance mechanisms. Friction is another problem in mechanical transmissions. The aspects which are most important to include in a dynamic simulation model of transmissions are (the text in this part of the section is to a large extent adopted from Hildre, H.P. (1991)):

- Nonlinear stiffness
- Backlash
- Friction
- Torque ripple (due to imperfection or from the motor)
- The input and output of the transmission must be represented in the mechanical simulation. This allows feedback measurement to the control module from both these systems
- Speed conversion from rotational movement to linear movement and/or a reduction ratio between two rotational movements

Backlash, nonlinear springs and friction can affect the performance significantly and must therefore be modeled precisely to achieve a simulation model close to the real system. The representation of the input and output of the transmission in the simulation model allows feedback measurements from both these systems. This feature also includes the dynamics in the transmission and the possibilities to add sensor dynamics.

Transmission systems can be designed in many ways and there are a lot of different types. The standard master-slave based transmission elements cannot model the mechanical imperfection that a transmission may have. A flexible transmission is built up from joints, springs and some special routines to transmit motion and forces between the joints that constitute the flexible transmission model. The following section shows some typical transmission types and how these can be modeled. These examples can then be further combined and new configurations can be built. Note that the generation of torque ripples from the motor can be modeled in the control module.

Reduction Gear

Fig. 3.16a shows an electric motor and a reduction gear connected to a revolute joint in a structure. This is a common configuration in many types of mechanisms. Fig. 3.16b show an equivalent scheme and how this unit is represented in a simulation. This transmission is modeled by two revolute joints, an "input" joint and a "gear joint" which are coincident with the structures "link joint". Links i and $i+1$ are part of this structure.

The gear spring k_g can be of a nonlinear type with backlash capability and has to be modeled according to the spring characteristic on the output shaft on the reduction gear. Gear manufacturers usually state the stiffness measured on the output shaft when the input shaft is fixed. The backlash is measured with the same procedure as the output stiffness. The part of link i which connects the two joints together represents all the static components such as the gear housing, motor housing and sensor housing. The link named "input link" represents the high speed moving parts such as high speed gear components, motor rotor and moving sensor components. Mass and inertia for both these links are according to the real system.

The input joint and gear joint are not direct coupled in the stiffness and mass/inertia matrix in the simulation. This coupling is done by two implicit functions. One setting the stress-free spring angle on the gear joint θ_{g0} to θ_i/N, where N is the reduction ratio and θ_i is the input angle. ($N > 1$ for reduction gears.) The other implicit function returns the actual gear joint torque (feedback torque) to the input link, $M_i = M_{gk}/N$. M_{gk} is the measured spring torque in the gear joint. The system will then iterate until the feedback torque is equal to the input torque (equilibrium). The position, velocity and acceleration for the input and gear joint are standard output variables from the simulation and can be plotted or used as feedback measurements to the control module.

Friction is an importent aspect for accurate modeling of gear behavior, however, details of friction modeling are not covered in this book.

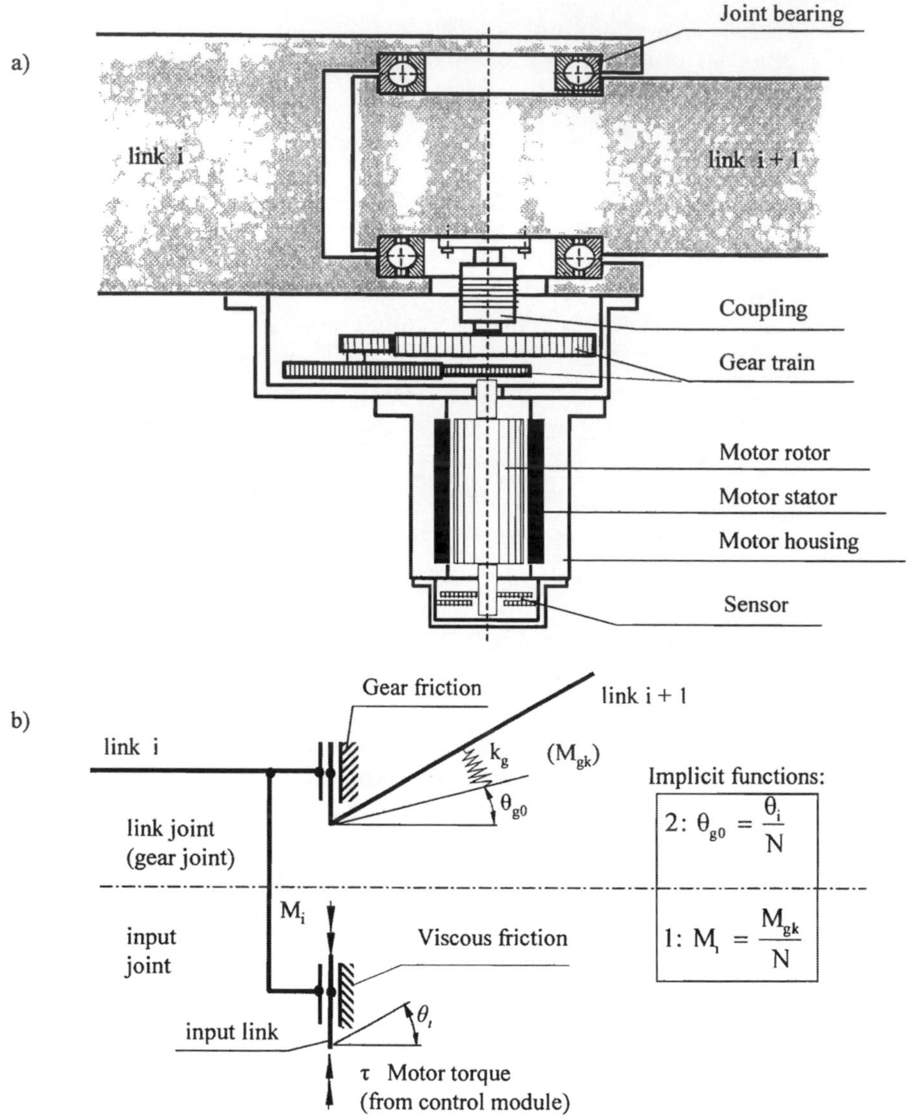

Figure 3.16: a) Electric motor and a reduction gear connected to a revolute joint. b) Equivalent simulation scheme.

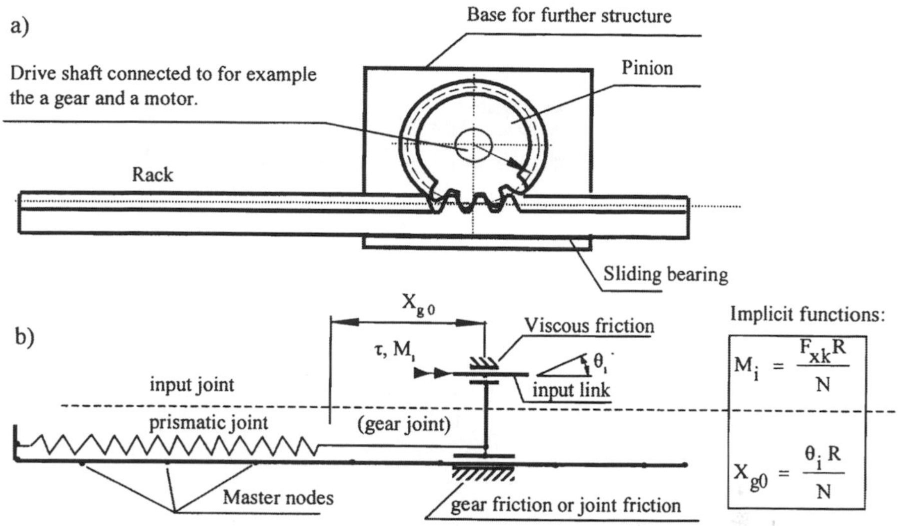

Figure 3.17: a) Rack and pinion. b) Equivalent simulation scheme.

Rack and Pinion

Fig. 3.17a shows a rack and pinion linear transmission. This type of transmission is for example used in gantry robots. Fig. 3.17b shows an equivalent scheme and how this unit is represented in the simulation. The unit is built by a rotary joint and a prismatic joint. The rotary joint representing the input drive and the gear joint which is coincident with the structure's prismatic joint representing the conversion to linear movement. The rotary drive can be modeled as a reduction gear. The linear spring can be of nonlinear type with backlash capability and has to be modeled according to the total spring stiffness of the transmission. The coupling between the rotary input joint and the linear movement is controlled by two implicit functions. One setting the stress-free spring length on the prismatic joint X_{g0} to $\theta_i R/N$, where N is the reduction ratio in the gear and R is the radius of the pinion. The other implicit function returns the actual prismatic spring force, F_{xk} to the input joint, $M_i = F_{xk}R/N$, where F_{xk} is the measured spring force.

It can be seen from Fig. 3.17 that the linear spring is connected to the end of the rack. The driving force will then also act in the end of the rack. In a real system these forces will act in the transmission unit (prismatic joint). The reason for this simplification is that the linear spring in the prismatic joint cannot change position with the nearest master node when the joint is moving. All other forces and bending in the joint will be carried by the

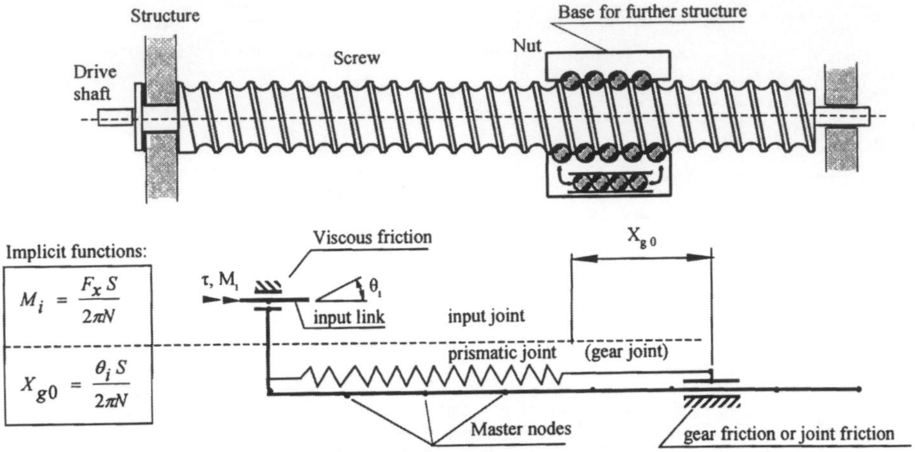

Figure 3.18: Ball screw and equivalent simulation scheme.

two nearest master nodes. This simplification will normally not affect the dynamics in the system. Note that frequencies for transversal vibrations are often several magnitudes lower than for axial vibrations.

The prismatic joint in the simulation cannot carry bending normal to the sliding direction and two prismatic joints should therefore be used if this option is required.

Ball Screw

Fig. 3.18a shows a ball screw transmission connected to a structure. This is a precision transmission widely used in all kinds of machines. Fig. 3.18b shows an equivalent scheme and how this unit is represented in the simulation. The unit is built by a rotary joint and a prismatic joint in combination. The rotary joint representing the input drive and screw and the gear joint which is coincident with the structure's prismatic joint representing the conversion to linear movement. The rotary drive can be modeled as a reduction gear. It should be noted that the inertia of the screw must be recalculated and added to the input joint. The linear spring can be of a nonlinear type with backlash capability and has to be modeled according to the spring stiffness of the transmission.

The coupling between the rotary input joint and linear transmission is controlled by two implicit functions. One setting the stress-free spring length on the prismatic joint X_{g0} to $\theta_i S/2\pi N$, where N is the reduction ratio and S is the screw pitch. The other implicit function returns the actual prismatic

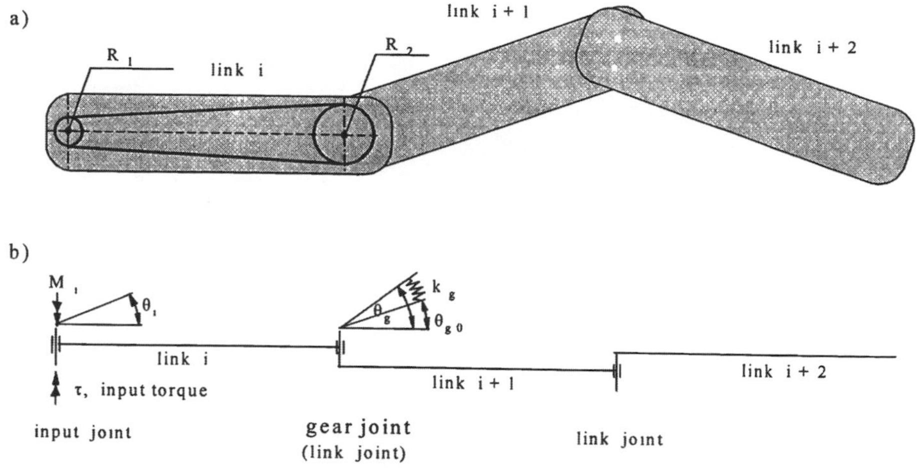

Figure 3.19: Simple belt transmission.

force to the input joint, $M_i = F_x S/2\pi N$, where F_x is the measured spring force.

Belt and Cable

Belt type transmissions are widely used, particularly in systems without relatively high performance requirements, this is due to low stiffness. On the other hand these systems have extremely low inertia and volume and are very cost effective. Flat steel belts have higher stiffness and some other advantages and are therefore also used in high performance systems.

Belt transmissions can be arranged in many ways and it is therefore difficult to describe general modeling techniques. Fig. 3.19 shows a belt transmission of links connected by a revolute joint. The belt is located in link i and drives link $i+1$. The belt stiffness is located at the gear joint which is coincident with the structure revolute joint. The belt is replaced by two implicit functions. One implicit function gives the stress-free spring angle as $\theta_{g0} = \theta_i R_1/R_2$ where θ_i is the angle of the input shaft. R_1 is the radius of the driving wheel at the input joint and R_2 is the radius at the driven gear wheel. The other implicit function returns the actual torque in the gear joint to the input joint, $M_i = M_{gk} R_1/R_2$ where M_{gk} is the measured spring torque in the gear joint. Forces in the joint due to preloading and time varying stresses in belts often have no practical effects on the dynamic performance. If this effect is important, a spring between the input joint and gear joint

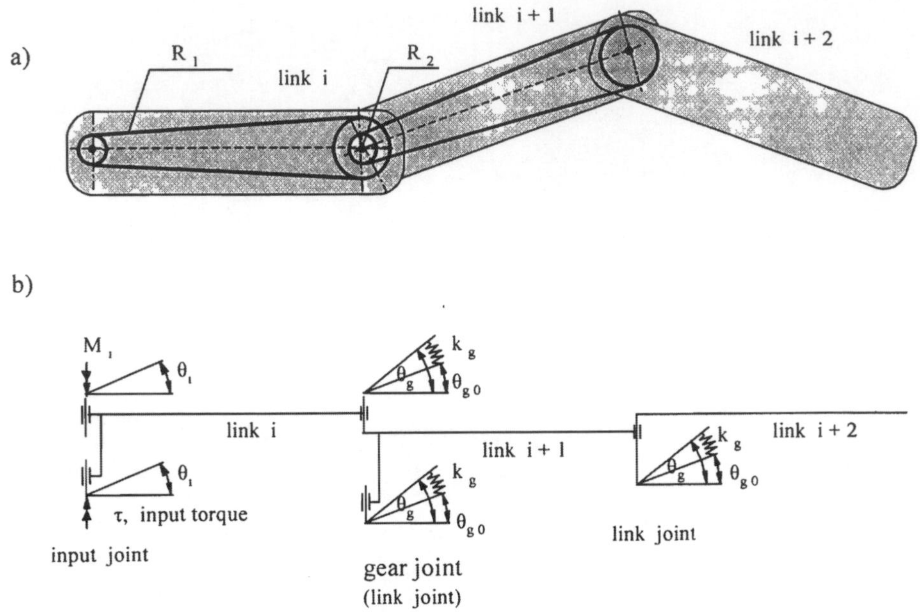

Figure 3.20: Concatenated belt transmission.

must be added. This spring force can be constant or variable controlled by an implicit function.

Assume that the structure above has an additional belt transmission. Link $i+2$ is driven by two belts, see Fig. 3.20. These belts are only controlling link $i+2$, but the angle of link $i+1$ will also affect the angle of the link $i+2$. This coupling can be done by locating the first gear joint in link $i+1$. The stiffness at the gear joint represents the stiffness of the belt in link i and the stiffness of the gear joint represents the stiffness of the belt in link $i+1$. These two belts are replaced by four implicit functions.

3.6.4 The Cam Joint

The cam surface is defined by a set of master triads, and these nodes are ordered in increasing sequence, see Fig. 3.21. The z-axis is pointing in the direction of the surface tangent, and the x-axis in the surface normal direction. If the first and last master are the same, the cam has a closed surface. One slave triad defines the cam follower. The local direction is irrelevant.

The cam joint is simulated by a nonlinear spring, which is connected between the cam curve and the cam follower. It is possible to connect a

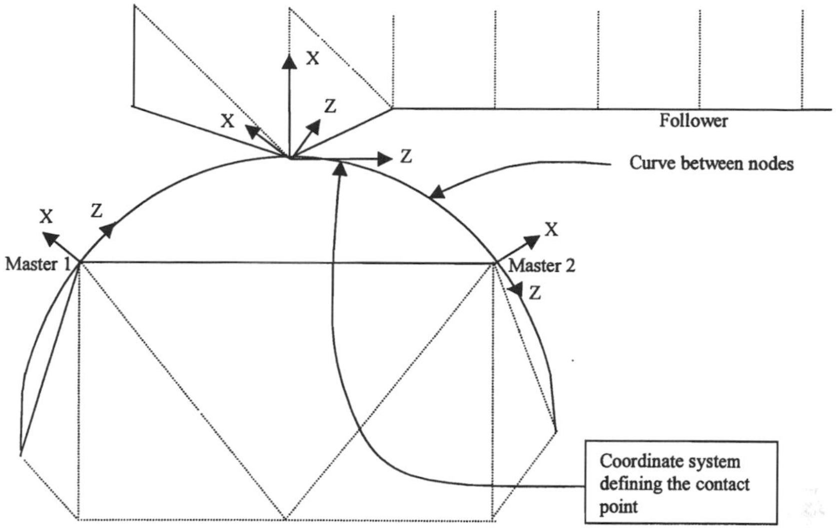

Figure 3.21: Coordinate system attached to the slave node and to the master nodes for the cam joint.

spring from the slave and the contact point in both the x- and y-directions. It is possible to add friction forces based on the forces from the x-spring.

The master triads define the surface of the cam. A new coordinate system is calculated at the contact point. This coordinate system has the z-axis in the cam surface direction and x-axis normal to the surface. The contact is enforced through nonlinear springs between masters and slaves. For details refer to Appendix B

3.7 Other Mechanism Elements

3.7.1 Additional Masses

Additional mass elements are used to add concentrated mass or inertia to specified DOFs of a mechanism model for instance to model flywheel effects or external bodies (payloads) to be handled by the mechanism. The additional masses specified are added directly to the diagonal of the system mass matrix for the specified DOFs. Moments of inertia will also introduce gyro effects, however, this is not discussed here.

3.7.2 Gravitational Forces

Gravitational forces are calculated from unit gravitational acceleration vectors in the x-, y- and z-directions referred to substructure local coordinate system, and denoted $\boldsymbol{U}_x$, $\boldsymbol{U}_y$ and $\boldsymbol{U}_z$, respectively. That is for all degrees of freedom in $\boldsymbol{U}_x$ corresponding to x-translation the acceleration component of $\boldsymbol{U}_x$ are set equal to 1, otherwise equal to 0. For $\boldsymbol{U}_y$ and $\boldsymbol{U}_z$ the acceleration components in the y- and z-directions, respectively, are set equal to 1, otherwise equal to 0.

The gravitational forces $\boldsymbol{G}_x$ corresponding to $\boldsymbol{U}_x$ are calculated from, see Eq. (3.20):

$$\begin{bmatrix} \boldsymbol{G}_{xi} \\ \boldsymbol{G}_{xe} \end{bmatrix} = \begin{bmatrix} \boldsymbol{M}_{ii} & \boldsymbol{M}_{ie} \\ \boldsymbol{M}_{ei} & \boldsymbol{M}_{ee} \end{bmatrix} \begin{bmatrix} \boldsymbol{U}_{xi} \\ \boldsymbol{U}_{xe} \end{bmatrix} = \begin{bmatrix} \boldsymbol{M}_{ii}\boldsymbol{U}_{xi} + \boldsymbol{M}_{ie}\boldsymbol{U}_{xe} \\ \boldsymbol{M}_{ei}\boldsymbol{U}_{xi} + \boldsymbol{M}_{ee}\boldsymbol{U}_{xe} \end{bmatrix} \tag{3.72}$$

The forces from unit gravitational acceleration in the x-direction are reduced by CMS-transformation to $\boldsymbol{g}_x$, refer to Eq. (3.21):

$$\begin{bmatrix} \boldsymbol{g}_{xe} \\ \boldsymbol{g}_{xg} \end{bmatrix} = \begin{bmatrix} \boldsymbol{I} & \boldsymbol{B}^T \\ \boldsymbol{0} & \boldsymbol{\Phi}^T \end{bmatrix} \begin{bmatrix} \boldsymbol{G}_{xe} \\ \boldsymbol{G}_{xi} \end{bmatrix} = \begin{bmatrix} \boldsymbol{G}_{xe} + \boldsymbol{B}^T\boldsymbol{G}_{xi} \\ \boldsymbol{\Phi}^T\boldsymbol{G}_{xi} \end{bmatrix} \tag{3.73}$$

Expanding by using Eq. (3.72) gives:

$$\boldsymbol{g}_{xe} = \boldsymbol{M}_{ie}^T\boldsymbol{U}_{xi} + \boldsymbol{M}_{ee}\boldsymbol{U}_{xe} + \boldsymbol{B}^T\boldsymbol{M}_{ii}\boldsymbol{U}_{xi} + \boldsymbol{B}^T\boldsymbol{M}_{ie}\boldsymbol{U}_{xe} \tag{3.74}$$

taking into account the symmetry property

$$\boldsymbol{M}_{ei} = \boldsymbol{M}_{ie}^T \tag{3.75}$$

and

$$\boldsymbol{g}_{xg} = \boldsymbol{\Phi}^T\boldsymbol{M}_{ii}\boldsymbol{U}_{xi} + \boldsymbol{\Phi}^T\boldsymbol{M}_{ie}\boldsymbol{U}_{xe} \tag{3.76}$$

The corresponding reduced gravitational forces from unit gravitation in the y-direction are easily found by change of indexes:

$$\boldsymbol{g}_{ye} = \boldsymbol{M}_{ie}^T\boldsymbol{U}_{yi} + \boldsymbol{M}_{ee}\boldsymbol{U}_{ye} + \boldsymbol{B}^T\boldsymbol{M}_{ii}\boldsymbol{U}_{yi} + \boldsymbol{B}^T\boldsymbol{M}_{ie}\boldsymbol{U}_{ye} \tag{3.77}$$

$$\boldsymbol{g}_{yg} = \boldsymbol{\Phi}^T\boldsymbol{M}_{ii}\boldsymbol{U}_{yi} + \boldsymbol{\Phi}^T\boldsymbol{M}_{ie}\boldsymbol{U}_{ye} \tag{3.78}$$

and similarly in the z-direction

$$\boldsymbol{g}_{ze} = \boldsymbol{M}_{ie}^T\boldsymbol{U}_{zi} + \boldsymbol{M}_{ee}\boldsymbol{U}_{ze} + \boldsymbol{B}^T\boldsymbol{M}_{ii}\boldsymbol{U}_{zi} + \boldsymbol{B}^T\boldsymbol{M}_{ie}\boldsymbol{U}_{ze} \tag{3.79}$$

$$\boldsymbol{g}_{zg} = \boldsymbol{\Phi}^T\boldsymbol{M}_{ii}\boldsymbol{U}_{zi} + \boldsymbol{\Phi}^T\boldsymbol{M}_{ie}\boldsymbol{U}_{ze} \tag{3.80}$$

During simulation the super element no. i may have a position specified by the SECCS called $\boldsymbol{t}_{SEi}$, see Fig. 3.10. Let us denote the direction cosine part of this global transformation matrix $\boldsymbol{R}_{SEi}$. If the global gravitational vector is denoted $\boldsymbol{V}$, the gravitational components $\boldsymbol{V}_{ci}$ in the actual super element i directions are calculated from:

$$\boldsymbol{V}_{ci} = \boldsymbol{R}_{SEi}^T \boldsymbol{V} \tag{3.81}$$

The super element gravitational forces $\boldsymbol{g}_i$ are now calculated from, see Eqs. (3.74 to (3.80):

$$\boldsymbol{g}_i = V_{ci}(1) \cdot \begin{bmatrix} g_{xe} \\ g_{xg} \end{bmatrix}_i + V_{ci}(2) \cdot \begin{bmatrix} g_{ye} \\ g_{yg} \end{bmatrix}_i + V_{ci}(3) \cdot \begin{bmatrix} g_{ze} \\ g_{zg} \end{bmatrix}_i \tag{3.82}$$

The gravitational forces are now transformed to the actual local super node direction, if any, and then added to the system force vector.

3.7.3 External Loading

External loading as concentrated forces and torques may be added to any system DOF as a constant or as a function of time. The loading may be added explicitly to a system DOF, or it may be added to a super node as a vector in space with a direction specified by two points. The position of these points may be constant if referred to the global coordinate system or moving if one or both points are referred to the coordinate system of a super element. That is the external loading may vary both in magnitude and direction during simulation of mechanism motion.

If an external loading is specified to an actual DOF at system level, its magnitude is calculated for the time position and added directly to the system force vector. If two points, $\boldsymbol{P}_{1i}$ and $\boldsymbol{P}_{2j}$ are specified relative to super elements i and j, respectively, the global positions of these two points are first calculated from

$$\begin{aligned} \boldsymbol{P}_1 &= \boldsymbol{t}_{SEi} \boldsymbol{P}_{1i} \\ \boldsymbol{P}_2 &= \boldsymbol{t}_{SEj} \boldsymbol{P}_{2j} \end{aligned} \tag{3.83}$$

The load vector $\boldsymbol{F}$ is now calculated from

$$\boldsymbol{F} = \frac{\boldsymbol{P}_2 - \boldsymbol{P}_1}{\|\boldsymbol{P}_2 - \boldsymbol{P}_1\|} \cdot f \tag{3.84}$$

where f is the magnitude of the loading for the actual time position. This external force $\boldsymbol{F}$ is added directly into the system vector to the super node specified, or if this super node has an updated local direction, $\boldsymbol{F}$ is transformed to this direction and then added into the system vector.

3.7.4 Prescribed Motion

Any DOF at system level may have a specified prescribed motion as a function of time. This is specified by the general constraint Eq. (3.50):

$$r_i = r_{prsc} = c_o + \sum_{j=1}^{P} c_j (r_{master})_j \tag{3.85}$$

where $P = 0$. The constraint equation may then be written

$$r_i = r_{prsc} = c_o = f(t) \tag{3.86}$$

where c_0 is the prescribed motion as a function of time.

3.7.5 Initial Velocities and Accelerations

As a default, initial velocities and accelerations are set to zero, that is the simulation starts from an equilibrium position with initially no motion. However, it may be of interest to simulate a steady state motion and then we should enter the actual velocities and/or accelerations for all system DOFs that have initial motion. In the general case these data may be found from a rigid-body kinematic analysis of the initial position.

An alternative approach for steady state motion simulation may be to start from an equilibrium position and then accelerate to the actual steady state motion.

Chapter 4

Numerical Simulation

4.1 Dynamic Simulation

4.1.1 Dynamic Equation on Incremental form

The *dynamic equation of motion* at time t may be written:

$$\boldsymbol{F}\left(t, \boldsymbol{r}, \dot{\boldsymbol{r}}, \ddot{\boldsymbol{r}}\right) = 0 \tag{4.1}$$

or

$$\boldsymbol{F}^I\left(t, \boldsymbol{r}, \dot{\boldsymbol{r}}, \ddot{\boldsymbol{r}}\right) + \boldsymbol{F}^D\left(t, \boldsymbol{r}, \dot{\boldsymbol{r}}, \ddot{\boldsymbol{r}}\right) + \boldsymbol{F}^S\left(t, \boldsymbol{r}, \dot{\boldsymbol{r}}, \ddot{\boldsymbol{r}}\right) - \boldsymbol{Q}\left(t, \boldsymbol{r}, \dot{\boldsymbol{r}}, \ddot{\boldsymbol{r}}\right) = 0 \tag{4.2}$$

where $\boldsymbol{F}^I$ represents inertia forces, $\boldsymbol{F}^D$ represents damping forces and $\boldsymbol{F}^S$ represents elastic forces. $\boldsymbol{Q}$ represents external forces. We like to integrate this equation in time with a time step length of h. At time t_k ($t_k = kh$) the equation of motion (Eq. 4.2) may be written as:

$$\boldsymbol{F}_k^I + \boldsymbol{F}_k^D + \boldsymbol{F}_k^S = \boldsymbol{Q}_k \tag{4.3}$$

and at time t_{k+1} the same equation may be written:

$$\boldsymbol{F}_{k+1}^I + \boldsymbol{F}_{k+1}^D + \boldsymbol{F}_{k+1}^S = \boldsymbol{Q}_{k+1} \tag{4.4}$$

By subtracting Eq. (4.3) from Eq. (4.4) we get:

$$\left(\boldsymbol{F}_{k+1}^I - \boldsymbol{F}_k^I\right) + \left(\boldsymbol{F}_{k+1}^D - \boldsymbol{F}_k^D\right) + \left(\boldsymbol{F}_{k+1}^S - \boldsymbol{F}_k^S\right) = \boldsymbol{Q}_{k+1} - \boldsymbol{Q}_k \tag{4.5}$$

or

$$\Delta\boldsymbol{F}_k^I + \Delta\boldsymbol{F}_k^D + \Delta\boldsymbol{F}_k^S = \Delta\boldsymbol{Q}_k \tag{4.6}$$

This is the equation of motion on incremental form. In the following, the different factors of Eq. (4.6) will be expanded.

The inertia, damping and stiffness relations are approximated by a linearization around the starting position for each increment, and the incremental system matrices, shown in the following, are generated for that position. The error introduced by this approximation may be eliminated by equilibrium iterations. The exact incremental system matrices, called the secant matrices, are a function of the unknown displacement increments sought for here, and cannot be generated in advance.

The incremental inertia forces may be written:

$$\Delta \boldsymbol{F}_k^I = \boldsymbol{M}_{Ik} \Delta \ddot{\boldsymbol{r}}_k \tag{4.7}$$

where $\boldsymbol{M}_{Ik}$ is the incremental system mass matrix at the beginning of time increment k and $\Delta \ddot{\boldsymbol{r}}_k = \ddot{\boldsymbol{r}}_{k+1} - \ddot{\boldsymbol{r}}_k$ represent the change in acceleration during increment k. The system mass matrix $\boldsymbol{M}_{Ik}$ may be constant or a function of the displacement vector $\boldsymbol{r}$, depending on what kind of element mass representation is used. The element mass matrices $\boldsymbol{m}$ are constant, but undergo a geometric transformation before they are added into the system matrix. If lumped mass representation is chosen, the element mass matrix is diagonal. The system mass matrix will then also be diagonal and constant during integration.

The incremental damping forces from Eq. (4.6) may be written:

$$\Delta \boldsymbol{F}_k^D = \boldsymbol{C}_{Ik} \Delta \dot{\boldsymbol{r}}_k \tag{4.8}$$

where $\boldsymbol{C}_{Ik}$ is the incremental system damping matrix at the beginning of time increment k and $\Delta \dot{\boldsymbol{r}}_k = \dot{\boldsymbol{r}}_{k+1} - \dot{\boldsymbol{r}}_k$ represent the change in velocity during increment k. For this formulation the system damping matrix may be constant or a function of the displacement vector $\boldsymbol{r}$.

The incremental elastic forces from Eq. (4.6) may be written

$$\Delta \boldsymbol{F}_k^S = \boldsymbol{K}_{Ik} \Delta \boldsymbol{r}_k \tag{4.9}$$

where $\boldsymbol{K}_{Ik}$ is the incremental system stiffness matrix at the beginning of time increment k and $\Delta \boldsymbol{r} = \boldsymbol{r}_{k+1} - \boldsymbol{r}_k$ represent the increment in displacement for increment k. In mechanisms, the system stiffness matrix is in general a function of the displacement vector $\boldsymbol{r}$.

The general form of the incremental (linearizied) dynamic equation of motion may now be written as follows.

$$\boldsymbol{M}_{Ik} \Delta \ddot{\boldsymbol{r}}_k + \boldsymbol{C}_{Ik} \Delta \dot{\boldsymbol{r}}_k + \boldsymbol{K}_{Ik} \Delta \boldsymbol{r}_k = \Delta \boldsymbol{Q}_k \tag{4.10}$$

In general $\boldsymbol{M}_I$, $\boldsymbol{C}_I$ and $\boldsymbol{K}_I$ are recalculated for each time increment and iteration. Solution of Eq. (4.10) by a time integration algorithm, for instance Newmark's method, gives $\Delta\boldsymbol{r}_k$, $\Delta\dot{\boldsymbol{r}}_k$ and $\Delta\ddot{\boldsymbol{r}}_k$. The total solution at the end of the increment is then:

$$\begin{aligned} \boldsymbol{r}_{k+1} &= \boldsymbol{r}_k + \Delta\boldsymbol{r}_k \\ \dot{\boldsymbol{r}}_{k+1} &= \dot{\boldsymbol{r}}_k + \Delta\dot{\boldsymbol{r}}_k \\ \ddot{\boldsymbol{r}}_{k+1} &= \ddot{\boldsymbol{r}}_k + \Delta\ddot{\boldsymbol{r}}_k \end{aligned} \tag{4.11}$$

The solution at the end of the increment (Eq. 4.11) may be used to calculate $\boldsymbol{F}^I_{k+1}$, $\boldsymbol{F}^D_{k+1}$ and $\boldsymbol{F}^S_{k+1}$, and because of the linearization there will be unbalanced forces at the end of the increment.

$$\Delta\hat{\boldsymbol{F}}_{k+1} = \boldsymbol{Q}_{k+1} - \left[\boldsymbol{F}^I_{k+1} + \boldsymbol{F}^D_{k+1} + \boldsymbol{F}^S_{k+1}\right] \tag{4.12}$$

These residual forces may be added to the load increment for the next step, see Eq. (4.6).

$$\Delta\boldsymbol{Q}_k = \boldsymbol{Q}_{k+1} - \boldsymbol{Q}_k + \Delta\hat{\boldsymbol{F}}_k = \boldsymbol{Q}_{k+1} - \left[\boldsymbol{F}^I_k + \boldsymbol{F}^D_k + \boldsymbol{F}^S_k\right] \tag{4.13}$$

Eq.(4.10) may then be written:

$$\boldsymbol{M}_{Ik}\Delta\ddot{\boldsymbol{r}}_k + \boldsymbol{C}_{Ik}\Delta\dot{\boldsymbol{r}}_k + \boldsymbol{K}_{Ik}\Delta\boldsymbol{r}_k = \boldsymbol{Q}_{k+1} - \left[\boldsymbol{F}^I_k + \boldsymbol{F}^D_k + \boldsymbol{F}^S_k\right] \tag{4.14}$$

This is an approximation for the equation of equilibrium at time t_{k+1}. To achieve equilibrium at the end of the increment, in the nonlinear case, iteration has to be used to minimize the error from the solution of Eq. (4.14). Iteration is accomplished by replacing $\Delta\boldsymbol{Q}_k$ in Eq. (4.10) by $\Delta\hat{\boldsymbol{F}}_{k+1}$ and by solving for the correction $\boldsymbol{\Delta}_k$ for $\Delta\boldsymbol{r}_k$ from:

$$\begin{aligned} {}^i\boldsymbol{M}_{Ik}{}^i\ddot{\boldsymbol{\Delta}}_k &+ {}^i\boldsymbol{C}_{Ik}{}^i\dot{\boldsymbol{\Delta}}_k + {}^i\boldsymbol{K}_{Ik}{}^i\boldsymbol{\Delta}_k \\ &= {}^{i-1}\boldsymbol{Q}_{k+1} - \left[{}^{i-1}\boldsymbol{F}^I_{k+1} + {}^{i-1}\boldsymbol{F}^D_{k+1} + {}^{i-1}\boldsymbol{F}^S_{k+1}\right] \end{aligned} \tag{4.15}$$

The super index to the left of the symbols indicates the iteration number. The displacement increment is then improved from:

$${}^i\Delta\boldsymbol{r}_k = {}^{i-1}\Delta\boldsymbol{r}_k + {}^i\boldsymbol{\Delta}_k \tag{4.16}$$

and the total displacement vector is calculated from

$${}^i\boldsymbol{r}_{k+1} = {}^{i-1}\boldsymbol{r}_{k+1} + {}^i\boldsymbol{\Delta}_k \tag{4.17}$$

Velocities and accelerations are calculated in the same way. Then ${}^i\boldsymbol{F}^I_{k+1}$, ${}^i\boldsymbol{F}^D_{k+1}$ and ${}^i\boldsymbol{F}^S_{k+1}$ may be calculated and substituted into Eq. (4.15) to solve for the correction ${}^{i+1}\boldsymbol{\Delta}_k$, and so on.

If $\boldsymbol{M}_{Ik}$, $\boldsymbol{C}_{Ik}$ and $\boldsymbol{K}_{Ik}$ are updated after each iteration, the process is called Newton-Raphson iteration. If the matrices $\boldsymbol{M}_{Ik}$, $\boldsymbol{C}_{Ik}$ and $\boldsymbol{K}_{Ik}$ are all left constant during iteration or only updated after some iterations, it is called modified Newton-Raphson iteration.

Some criteria of iteration convergence must be established. A convergence test could be

$$\left\| {}^{i}\boldsymbol{\Delta}_k \right\| \leq \varepsilon \tag{4.18}$$

where ε is a specified tolerance.

4.1.2 The Newmark Integration Algorithm

The Newmark's β-family of algorithms may be used for time integration, and the base for this method is:

$$\begin{aligned} \dot{\boldsymbol{r}}_{k+1} &= \dot{\boldsymbol{r}}_k + (1-\gamma)\, h\ddot{\boldsymbol{r}}_k + \gamma h \ddot{\boldsymbol{r}}_{k+1} \\ \boldsymbol{r}_{k+1} &= \boldsymbol{r}_k + h\dot{\boldsymbol{r}}_k + \left(\frac{1}{2} - \beta\right) h^2 \ddot{\boldsymbol{r}}_k + \beta h^2 \ddot{\boldsymbol{r}}_{k+1} \end{aligned} \tag{4.19}$$

Where β and γ are integration parameters, and h is the time increment. These equations may be rewritten as

$$\Delta\dot{\boldsymbol{r}}_k = h\ddot{\boldsymbol{r}}_k + \gamma h \Delta\ddot{\boldsymbol{r}}_k \tag{4.20}$$

$$\Delta\boldsymbol{r}_k = h\dot{\boldsymbol{r}}_k + \frac{1}{2}h^2\ddot{\boldsymbol{r}}_k + \beta h^2 \Delta\ddot{\boldsymbol{r}}_k \tag{4.21}$$

Now we like to express the increment in velocity and acceleration as a function of the increment in displacement and by known parameters at time t_k. Eq. (4.21) gives directly:

$$\Delta\ddot{\boldsymbol{r}}_k = \frac{1}{\beta h^2}\Delta\boldsymbol{r}_k - \frac{1}{\beta h}\dot{\boldsymbol{r}}_k - \frac{1}{2\beta}\ddot{\boldsymbol{r}}_k = \frac{1}{\beta h^2}\Delta\boldsymbol{r}_k - \boldsymbol{a}_k \tag{4.22}$$

where

$$\boldsymbol{a}_k = \frac{1}{\beta h}\dot{\boldsymbol{r}}_k + \frac{1}{2\beta}\ddot{\boldsymbol{r}}_k \tag{4.23}$$

Combining Eqs. (4.22) and (4.20) gives:

$$\Delta\dot{\boldsymbol{r}}_k = \frac{\gamma}{\beta h}\Delta\boldsymbol{r}_k - \frac{\gamma}{\beta}\dot{\boldsymbol{r}}_k - \left[\frac{\gamma h}{2\beta} - h\right]\ddot{\boldsymbol{r}}_k = \frac{\gamma}{\beta h}\Delta\boldsymbol{r}_k - \boldsymbol{b}_k \tag{4.24}$$

where

$$\boldsymbol{b}_k = \frac{\gamma}{\beta}\dot{\boldsymbol{r}}_k + \left[\frac{\gamma}{2\beta} - 1\right] h\ddot{\boldsymbol{r}}_k \tag{4.25}$$

Eqs. (4.22) and (4.24) into Eq. (4.14) gives:

$$\boldsymbol{N}_{Ik}\Delta\boldsymbol{r}_k = \Delta\hat{\boldsymbol{Q}}_k \tag{4.26}$$

where the Newton matrix is:

$$\boldsymbol{N}_{Ik} = \boldsymbol{K}_{Ik} + \frac{\gamma}{\beta h}\boldsymbol{C}_{Ik} + \frac{1}{\beta h^2}\boldsymbol{M}_{Ik} \tag{4.27}$$

and

$$\Delta\hat{\boldsymbol{Q}}_k = \boldsymbol{Q}_{k+1} - \left[\boldsymbol{F}_k^I + \boldsymbol{F}_k^D + \boldsymbol{F}_k^S\right] + \boldsymbol{C}_{Ik}\boldsymbol{b}_k + \boldsymbol{M}_{Ik}\boldsymbol{a}_k \tag{4.28}$$

$\boldsymbol{Q}_{k+1}$ represents external loading at the end of the time increment, which is the input forces and torques supplied to the mechanism. Evaluation of the Newton matrix $\boldsymbol{N}_k$ and of the internal inertia, damping and stiffness forces $\boldsymbol{F}^I$, $\boldsymbol{F}^D$ and $\boldsymbol{F}^S$, respectively, are covered later in this section.

Possible equilibrium iterations are based on Eq. (4.15). The improved displacement increment from Eq. (4.16) into Eqs. (4.22) and (4.24), respectively, gives:

$$^{i}\Delta\ddot{\boldsymbol{r}}_k = {}^{i-1}\Delta\ddot{\boldsymbol{r}}_k + {}^{i}\ddot{\boldsymbol{\Delta}}_k = \frac{1}{\beta h^2}\left({}^{i-1}\Delta\boldsymbol{r}_k + {}^{i}\boldsymbol{\Delta}_k\right) - \boldsymbol{a}_k \tag{4.29}$$

$$^{i}\Delta\dot{\boldsymbol{r}}_k = {}^{i-1}\Delta\dot{\boldsymbol{r}}_k + {}^{i}\dot{\boldsymbol{\Delta}}_k = \frac{\gamma}{\beta h}\left({}^{i-1}\Delta\boldsymbol{r}_k + {}^{i}\boldsymbol{\Delta}_k\right) - \boldsymbol{b}_k \tag{4.30}$$

and from this it follows that

$$\begin{aligned} {}^{i}\ddot{\boldsymbol{\Delta}}_k &= \frac{1}{\beta h^2}{}^{i}\boldsymbol{\Delta}_k \\ {}^{i}\dot{\boldsymbol{\Delta}}_k &= \frac{\gamma}{\beta h}{}^{i}\boldsymbol{\Delta}_k \end{aligned} \tag{4.31}$$

Combining Eq. (4.31) with Eq. (4.15) gives:

$$^{i}\boldsymbol{N}_{Ik}{}^{i}\boldsymbol{\Delta}_k = {}^{i-1}\Delta\hat{\boldsymbol{F}}_k \tag{4.32}$$

where

$$^{i}\boldsymbol{N}_{Ik} = {}^{i}\boldsymbol{K}_{Ik} + \left(\frac{\gamma}{\beta h}\right){}^{i}\boldsymbol{C}_{Ik} + \left(\frac{1}{\beta h^2}\right){}^{i}\boldsymbol{M}_{Ik} \tag{4.33}$$

and

$$^{i-1}\Delta\hat{\boldsymbol{F}}_k = {}^{i-1}\boldsymbol{Q}_{k+1} - \left[{}^{i-1}\boldsymbol{F}^I_{k+1} + {}^{i-1}\boldsymbol{F}^D_{k+1} + {}^{i-1}\boldsymbol{F}^S_{k+1}\right] \tag{4.34}$$

Here $(i-1)$ refers to values calculated for system variables from previous iteration.

If $\boldsymbol{K}_{Ik}$, $\boldsymbol{C}_{Ik}$ and $\boldsymbol{M}_{Ik}$ are left constant during some or all iterations, as for modified Newton-Raphson, the effective stiffness matrix in Eq. (4.32) is unchanged for the same iterations and new triangularizations are avoided. The iteration continues until some convergence test is satisfied, refer to Eq. (4.18).

For the Newmark β-family of integration algorithms, γ and β are integration parameters that are selected for controlling stability, accuracy and efficiency for the integration.

For the linear case it may be shown that the method is unconditionally stable for:

$$\begin{aligned} \gamma &\geq \frac{1}{2} \\ \beta &\geq \frac{1}{4}\left(\gamma + \frac{1}{2}\right)^2 \end{aligned} \tag{4.35}$$

For smaller β-values the method is only conditionally stable. The stability criteria may then be shown to be

$$h_{cr} = \frac{T}{2\pi}\left(\frac{1}{4}\left(\gamma + \frac{1}{2}\right)^2 - \beta\right)^{-\frac{1}{2}} \tag{4.36}$$

where h_{cr} is the critical time increment, T is the period for the highest frequency in the model and π is 3.14159...

The parameter γ may be selected to introduce artificial damping into the integration process. $\gamma > 1/2$ gives positive artificial damping, that is the amplitude will decay with increasing k. $\gamma < 1/2$ gives negative artificial damping, that is the amplitude will increase with increasing k. $\gamma = 1/2$ gives no artificial damping. Consequently, $\gamma = 1/2$ is often the selection. In this case, the Newton β-family includes the following methods:

$\beta = 0$: The second central difference method with $h_{cr} = 0,318T$

$\beta = 1/12$: Fox-Goodwin's method with $h_{cr} = 0,389T$

$\beta = 1/6$: Linear acceleration with $h_{cr} = 0,551T$

$\beta = 1/4$: Constant average acceleration (trapezoid method) that is unconditionally stable for linear systems.

$\beta = 1/4$ is very often the selection.

4.1.3 Evaluation of the Newton Matrix

The Newton matrix is a function of the integration parameters γ, β and of the integration time increment h, see Eq. (4.27). In Section 3.2, the mass and stiffness matrices are developed for each of the substructures and reduced into super element matrices by the CMS reduction techniques. However, the damping matrix in Eq. (3.20) and Eq. (3.21), is not developed. If we make the assumption that the damping force is proportional to the velocity of each mass point we will have:

$$\boldsymbol{C} = \alpha_1 \boldsymbol{M}$$

where α_1 is a constant. Similarly, if we assume the damping force proportional to the strain velocity in each point we will have:

$$\boldsymbol{C} = \alpha_2 \boldsymbol{K}$$

where α_2 is a constant. Combining these two assumptions we will have what is called *Rayleigh-damping* or *proportional damping.*

$$\boldsymbol{C} = \alpha_1 \boldsymbol{M} + \alpha_2 \boldsymbol{K} \tag{4.37}$$

It may now be shown that the damping ratio for the different natural frequencies may be calculated from:

$$\lambda_i = \frac{1}{2}\left(\frac{\alpha_1}{\omega_i} + \alpha_2 \omega_i\right) \tag{4.38}$$

where α_1 damps out lower vibration modes while α_2 damps out higher modes. If the damping ratios λ_i for two vibration modes are selected, the corresponding constants of proportionality, α_1 and α_2 may be calculated from

$$\begin{aligned} \alpha_1 &= \frac{2\omega_1\omega_2}{\omega_2^2 - \omega_1^2}\left(\lambda_1\omega_2 - \lambda_2\omega_1\right) \\ \alpha_2 &= \frac{2\left(\omega_2\lambda_2 - \omega_1\lambda_1\right)}{\omega_2^2 - \omega_1^2} \end{aligned} \tag{4.39}$$

where ω_1 and ω_2 are the circle frequencies and λ_1 and λ_2 are the damping ratios for the selected vibration modes, see figure Fig. 4.1

We may now calculate a super element Newton matrix from the relation, see Eqs. (4.27) and (4.37).

$$\boldsymbol{n}_i = \boldsymbol{k}_i + \frac{\gamma}{\beta h}\left(\alpha_1 \boldsymbol{m}_i + \alpha_2 \boldsymbol{k}_i\right) + \frac{1}{\beta h^2}\boldsymbol{m}_i \tag{4.40}$$

where

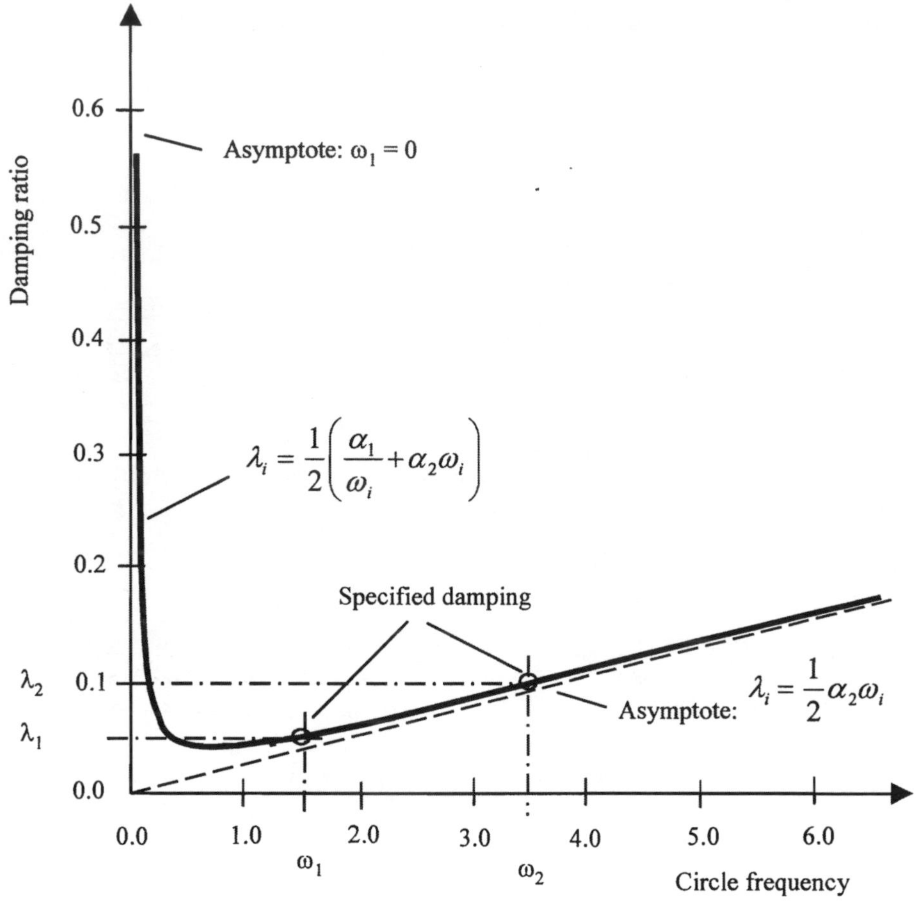

Figure 4.1: Typical relationship between damping and natural frequency arising from the specification of damping ratio at the frequencies.

$\boldsymbol{n}_i$: reduced super element Newton matrix

$\boldsymbol{k}_i$: reduced super element stiffness matrix

$\boldsymbol{m}_i$: reduced super element mass matrix

The integration parameters γ and β, and the constants of proportional damping α_1 and α_2 are usually constant during a simulation, and the super elements Newton matrices therefore only need to be updated each time the integration time increment h is changed.

Each time the integration algorithm requires a new Newton system matrix, see Eqs. (4.26) and (4.32), the super element Newton matrices are transformed to the actual directions of the DOFs at system level by

$$\bar{\boldsymbol{n}}_i = \boldsymbol{T}_{SEi}^T \boldsymbol{n}_i \boldsymbol{T}_{SEi} \tag{4.41}$$

(see Eqs. (3.43) to (3.47)). Here $\bar{\boldsymbol{n}}_i$ is the geometrical transformed super element Newton matrix. The super element Newton matrices are added into the incremental Newton matrix at system level as indicated by

$$\boldsymbol{N}_{Ik} = \sum_i \boldsymbol{a}_i^T \bar{\boldsymbol{n}}_i \boldsymbol{a}_i \tag{4.42}$$

where $\boldsymbol{a}_i$ are incidence matrices that represent super element topology at system level, see also Eqs. (3.48) and (3.49).

The stiffness for a spring may be a constant, or in the nonlinear case a variable dependent on the actual spring deflection. For axial springs the stiffness is transformed to the actual directions for the connected super nodes and then added into the Newton system matrix. Stiffness for joint springs are added directly to the diagonal of the involved DOFs of the system Newton matrix and to their coupling elements.

The damping coefficient of a damper may be a constant, or in the nonlinear case, a variable dependent on the actual damper velocity. For the Newton matrix, the damping coefficient must be multiplied by the factor $\gamma/\beta h$, refer to Eqs. (4.27) and (4.33). For axial dampers the modified damping coefficient is transformed to the actual directions for the connected super nodes and then added into the Newton system matrix. The modified damping coefficients for joint dampers are added to the diagonal of the involved DOFs of the Newton system matrix and to their coupling elements.

Additional masses may be regarded as constants during simulation. For the Newton matrix, additional masses must be multiplied by the factor $1/\beta h^2$, see Eqs. (4.27) and (4.30). The modified additional masses are added to the diagonal elements for the specified DOFs of the Newton matrix. Refer also to Section 3.7.

4.1.4 Evaluation of the Force Vector

The force vector must be evaluated for each time increment and iteration, see Eqs. (4.28) and (4.34). The external forces at time increment $k+1$, $\boldsymbol{Q}_{k+1}$, is calculated from:

- Super element gravitational forces
- Concentrated external forces in super nodes
- Gravitational forces from additional masses

The gravitational forces are calculated in Eq. (3.82). They are transformed to actual directions of the DOFs at system level by the use of the transformation matrix of Eq. (3.45):

$$\overline{\boldsymbol{g}}_i = \boldsymbol{T}_{SEi}^T \boldsymbol{g}_i \tag{4.43}$$

See also Eqs. (3.46) and (3.47). The transformed super element gravitational forces are added into the force vector by

$$\boldsymbol{Q}_{k+1} = \boldsymbol{Q}_{k+1} + \sum_i \boldsymbol{a}_i^T \overline{\boldsymbol{g}}_i \tag{4.44}$$

(see Eqs. (3.48) and (3.49)).

Direction and magnitude for specified concentrated loading are calculated for the actual position of the mechanism, and then transformed to the actual super node direction by the transformation matrix from Eq. (3.43) or Eq. (3.44) and added into the system vector $\boldsymbol{Q}_{k+1}$. For loading on specified DOFs at system level, the magnitude for the actual position is added directly into $\boldsymbol{Q}_{k+1}$.

For additional masses added to translational DOFs, a gravitational force is calculated from multiplying the component of the gravitational vector along the actual DOF by the specified mass. These gravitational forces are then added to $\boldsymbol{Q}_{k+1}$ on the corresponding DOFs.

If we now look at Eq. (4.28) and take into account that we may express

$$\boldsymbol{F}_k^I = \boldsymbol{M}_{Ik} \ddot{\boldsymbol{r}}_k \tag{4.45}$$

$$\boldsymbol{F}_k^D = \boldsymbol{C}_{Ik} \dot{\boldsymbol{r}}_k \tag{4.46}$$

this may be used to rewrite Eq. (4.28) to:

$$\begin{aligned} \Delta\hat{\boldsymbol{Q}}_k &= \boldsymbol{Q}_{k+1} + \boldsymbol{C}_{Ik}\left[\left(\frac{\gamma}{\beta}-1\right)\dot{\boldsymbol{r}}_k + \left(\frac{\gamma}{2\beta}-1\right)h\ddot{\boldsymbol{r}}_k\right] \\ &\quad + \boldsymbol{M}_{Ik}\left[\frac{1}{\beta h}\dot{\boldsymbol{r}}_k + \left(\frac{1}{2\beta}-1\right)\ddot{\boldsymbol{r}}_k\right] - \left[\boldsymbol{F}_k^S\right] \\ &= \boldsymbol{Q}_{k+1} + \boldsymbol{C}_{Ik}\boldsymbol{d}_k + \boldsymbol{M}_{Ik}\boldsymbol{a}_k - \boldsymbol{F}_k^S \end{aligned} \tag{4.47}$$

where

$$d_k = \left(\frac{\gamma}{\beta} - 1\right)\dot{r}_k + \left(\frac{\gamma}{2\beta} - 1\right)h\ddot{r}_k \tag{4.48}$$

and

$$a_k = \frac{1}{\beta h}\dot{r}_k + \left(\frac{1}{2\beta} - 1\right)\ddot{r}_k \tag{4.49}$$

For the first iteration within each time increment, the effective inertia forces are calculated by exchanging $\ddot{r}_k$ by $-a_k$ in Eq. (4.45) and $\dot{r}_k$ by $-d_k$ in Eq. (4.46). To avoid assembling the system mass and damping matrix explicitly, the corresponding super element accelerations and velocities are extracted from the system vectors and transformed to local super element direction. The local super element mass and damping matrices are then multiplied by the correspondning local acceleration and velocity vectors to produce the super element inertia and damping forces, respectively. These forces are then transformed to the actual super node directions and added into the system vector of internal forces, see Eqs. (4.43) and (4.44).

The super element stiffness forces are calculated as shown in Eqs. (3.39) to (3.42) in Section 3.3. These forces are then transformed to actual direction of DOFs for super element i,

$$\bar{S}_i = T_{SEi}^T S_i \tag{4.50}$$

see Eq. (4.43), and formally added into the force vector by

$$F_i^S = \sum_i a_i^T \bar{S}_i \tag{4.51}$$

see Eq. (4.44).

In the general case, a spring may have a stress-free length that is a variable of time, and it may also have a spring coefficient that is a variable of the spring deflection. At a certain position of the mechanism, the specified stress-free length of the spring l_o is evaluated and subtracted from the actual spring length l to give the spring deflection

$$d = l(r) - l_o(t, r) \tag{4.52}$$

In the general nonlinear case, the spring force is found by integration of the stiffness coefficient over the deflection. For axial springs, the evaluated spring force is transformed to the actual direction of the end super nodes and added into the force vector in a similar way as in Eqs. (4.50) and (4.51).

For joint springs, the torque or force is calculated in a similar way as for axial springs, however, the torque or force is added directly to the involved DOFs of the load vector without any geometric transformation.

A damper will, in the general case, have a coefficient that is a function of the damper velocity. For an actual damper velocity, the damping coefficient and damper torque or force is calculated. Addition into the load vector for the axial and joint dampers is done in the same way as described above for the corresponding spring elements.

The inertia forces or torques from additional masses on specified DOFs of the mechanism are calculated by multiplying the actual DOF acceleration by the magnitude of the actual mass. These forces or torques are then added to the corresponding DOFs of the load vector.

4.1.5 Quasistatic Equilibrium

Dynamic simulation of mechanism motion should start from an equilibrium position, otherwise unbalanced forces in the mechanism from the first time increment of the simulation may cause a false super imposed vibration in the simulation results. This may be the case both when the simulation starts from a stationary position or from a position with initial velocities and acceleration. Unbalanced forces for the stationary position will come from gravitation forces, initial loaded springs and from initial external loading, but also from positioning inaccuracies from the mechanism modeling. Equilibrium for initial positions of mechanisms with initial velocities and accelerations will also include unbalanced inertia and damping forces as well in the initial position. The inclusion of possible inertia and damping effects in the initial equilibrium iteration justifies the term *quasistatic equilibrium* iteration. Let us rewrite Eq. (4.15).

$$\begin{aligned} {}^{i}\boldsymbol{M}_{Ik}{}^{i}\ddot{\boldsymbol{\Delta}}_k &+ {}^{i}\boldsymbol{C}_{Ik}{}^{i}\dot{\boldsymbol{\Delta}}_k + {}^{i}\boldsymbol{K}_{Ik}{}^{i}\boldsymbol{\Delta}_k \\ &= {}^{i-1}\boldsymbol{Q}_{k+1} - \left[{}^{i-1}\boldsymbol{F}^{I}_{k+1} + {}^{i-1}\boldsymbol{F}^{D}_{k+1} + {}^{i-1}\boldsymbol{F}^{S}_{k+1}\right] \end{aligned} \tag{4.53}$$

However, for the quasistatic equilibrium iteration the initial velocities and accelerations are zero or specified and therefore by definition the corrections in the inertia and damping terms on the left-hand side of the above equation will cancel out, and the quasistatic iteration equation will then be

$$\begin{aligned} {}^{i}\boldsymbol{K}_{I0}{}^{i}\boldsymbol{\Delta}_0 & \\ &= {}^{i-1}\boldsymbol{Q}_0 - \left[{}^{i-1}\boldsymbol{F}^{I}_{0} + {}^{i-1}\boldsymbol{F}^{D}_{0} + {}^{i-1}\boldsymbol{F}^{S}_{0}\right] \end{aligned} \tag{4.54}$$

Here the subindexes are changed to 0, indicating the initial position. If the rigid body mechanism motions are all controlled by springs, as in a car suspension system, the stiffness matrix will be nonsingular and Eq. (4.54) may be used directly to iterate for the equilibrium position.

However, mechanism motion controlled by force input, Eq. (4.54), may be singular and *additional boundary conditions* must be introduced into the equation to eliminate the rigid body DOFs of the mechanism. Extra boundary conditions for the quasistatic equilibrium iteration are then usually specified for DOFs where external forces are applied. This should modify Eq. (4.54) so it is nonsingular, and the iteration could continue until equilibrium is reached within some specified tolerance.

Equilibrium iteration procedure

Starting from Eq. (4.54), possibly modified by additional boundary conditions, the iteration will follow the same algorithm as the dynamic equilibrium iteration described earlier in this section, with the exception that the stiffness matrix $\boldsymbol{K}_{I0}$ replaces the Newton matrix from the dynamic case. For each iteration the initial position of the mechanism is improved from

$$^{i}\boldsymbol{r}_0 = {}^{i-1}\boldsymbol{r}_0 + {}^{i}\boldsymbol{\Delta}_0$$

The updating of the velocity and acceleration vector in the dynamic iteration procedure are skipped in the quasistatic case. The stiffness matrix is evaluated in the same way as the Newton matrix in Section 4.1.3. All the stiffness terms from super element matrices and springs are kept while the mass and damping terms from super element matrices, additional masses and dampers are skipped.

The external force vector $^{i-1}\boldsymbol{Q}_0$ in Eq. (4.54) are evaluated in the same way as for the dynamic case for the initial integration time. The inertia and damping forces,

$$^{i-1}\boldsymbol{F}_0^I \; and \; {}^{i-1}\boldsymbol{F}_0^D$$

from Eq. (4.54) are evaluated based on constant velocities and accelerations during iteration, however, these vectors may also change during iteration due to updated positions of the mechanism. The stiffness forces

$$^{i-1}\boldsymbol{F}_0^S$$

are evaluated in exactly the same way as for the dynamic case.

The iteration based on Eq. (4.54) is called Newton-Raphson iteration or modified Newton-Raphson if the stiffness matrix is kept constant during some or all iterations. If the stiffness matrix is kept constant during iteration, the last evaluated and triangularized stiffness matrix is kept for new iterations, see Section 2.4. Only the *residual force* vector, the right-hand side of Eq. (4.54), is then evaluated for each iteration, and the increments for improving the mechanism position are evaluated through backward substitution. When this can be utilized, and many iterations are necessary, a lot of computational effort may be saved.

4.1.6 Frequency Analysis

The natural frequencies and the corresponding vibration modes of a mechanical system such as a mechanism are often very important information regarding a new design, for instance when evaluating if the frequency content of external or internal loading may excite any vibration mode. For a strongly nonlinear system, like a mechanism with a change in geometric position during operation, the natural frequencies and corresponding vibration modes may change significantly for the different positions of the mechanism. The dynamic equation on incremental form may be written, see Eq. (4.10).

$$\boldsymbol{M}_{Ik}\Delta\ddot{\boldsymbol{r}}_k + \boldsymbol{C}_{Ik}\Delta\dot{\boldsymbol{r}}_k + \boldsymbol{K}_{Ik}\Delta\boldsymbol{r}_k = \Delta\boldsymbol{Q}_k \tag{4.55}$$

For free undamped vibration

$$\boldsymbol{C}_{Ik} = \boldsymbol{0} \quad and \quad \Delta\boldsymbol{Q}_k = \boldsymbol{0}$$

and the equation may be written:

$$\boldsymbol{M}_{Ik}\Delta\ddot{\boldsymbol{r}}_k + \boldsymbol{K}_{Ik}\Delta\boldsymbol{r}_k = \boldsymbol{0} \tag{4.56}$$

Assuming harmonic oscillation we may introduce

$$\Delta\boldsymbol{r} = \boldsymbol{\Phi}\sin\omega t \tag{4.57}$$

where

$\boldsymbol{\Phi}$: is the eigenvector

ω: is the circle frequency

t: is the time

Introducing Eq. (4.57) and its second derivative into Eq. (4.56) we have

$$\left(\boldsymbol{K}_{Ik} - \omega^2\boldsymbol{M}_{Ik}\right)\boldsymbol{\Phi} = \boldsymbol{0} \tag{4.58}$$

and this equation represents the general eigenvalue problem, see Eq. (2.97). A number of the smallest eigenvalues and corresponding eigenvectors are evaluated from Eq. (4.58), see also Section 2.4.

For the eigenvalue analysis the stiffness matrix must be evaluated as for the quasistatic equilibrium iteration, and in addition the mass matrix $\boldsymbol{M}_{Ik}$ is evaluated in a similar way. As for the quasistatic equilibrium iteration, if the stiffness matrix is singular, additional boundary conditions must be introduced.

4.2 Control in Mechanism Simulation

Mechanisms are often connected to or acted upon by other types of elements such as sensors, controllers and actuators. For simplicity such elements are referred to as control elements. A need for multidisciplinary mechanism/FE/control simulation can therefore clearly be seen (the text in this section is adopted from Iversen, T. (1990)).

4.2.1 Problem Statement

The basic problem is to develop and implement a control model package for simulation in a composite system. Separate numerical methods are used, meaning that only minor changes are imposed on the FE part.

The structure equations constitutes a 2nd order system, which in the linear case can be written as:

$$\boldsymbol{M}\ddot{\boldsymbol{r}} + \boldsymbol{C}\ddot{\boldsymbol{r}} + \boldsymbol{K}\boldsymbol{r} = \boldsymbol{Q}(t) \tag{4.59}$$

where $\boldsymbol{r}$ is the vector of displacement, $\boldsymbol{M}$, $\boldsymbol{C}$ and $\boldsymbol{K}$ are matrices for mass, damping and stiffness, respectively, and $\boldsymbol{Q}(t)$ a vector of time dependent forces acting on the structure.

The control equations can usually be brought into the following form:

$$\dot{\boldsymbol{x}} = \boldsymbol{f}(t, \boldsymbol{u}, \boldsymbol{x}, \boldsymbol{z}) \tag{4.60}$$

$$\boldsymbol{0} = \boldsymbol{g}(t, \boldsymbol{u}, \boldsymbol{x}, \boldsymbol{z}) \tag{4.61}$$

where $\boldsymbol{u}$, $\boldsymbol{x}$ and $\boldsymbol{z}$ are vectors of inputs, state variables and algebraic variables, respectively. Most control elements are not explicitly time dependent. The exceptions are elements with inherent clock functions, as in a sample and hold element or a multiplexing unit.

The two parts are coupled when some elements in the displacement vector enter into the input vector $\boldsymbol{u}$, while in response some of the forces in $\boldsymbol{Q}(t)$ are taken from the control variables $\boldsymbol{x}$ or $\boldsymbol{z}$. The remaining part of the input $\boldsymbol{u}$ could be time functions such as a controller reference.

The system for the solution of the structure part is well established. It is based on Newmark's β-method with $\gamma = 1/2$ and $\beta = 1/4$, see Section 4.1. These parameters correspond to the trapezoidal rule.

The discretized structural part usually has much higher dimensions than the control part and therefore represents the heavy part of the computations. From accuracy considerations it is expected that the structural computations will limit the time step more than the control part. However, there might be cases where the coupling itself between the two parts limits the time step more than any of the two parts.

A subroutine package with a pre-defined interface between the control and the FE part is developed, which essentially integrates the variables in the differential-algebraic system one step forward from a given state. The solution for the control part is iterated until convergence on each call, and these iterations therefore form an inner loop of the iterations in the FE part.

4.2.2 Data Structure

The system will have to divide between the three types of variables: inputs, state variables and algebraic variables. The user must label the inputs while the system automatically determines the other two types. The inputs are either external time functions like a controller reference, or outputs from FE part that should not be changed in the control system. All types of variables appear in the same vector, and with the use of labels, there will be no need for rearrangement.

Thereby the variables for each module in a configuration can be stored successively as specified by the user. Each type of control module is identified through a type number and has a number of inputs, internal states and outputs. The internal states need not be connected to other modules. In addition, there is a number of parameters to be given by the user. Fig. 4.2 shows a module of type nn, with i inputs, j internal states, k outputs and m parameters.

There are two remarks about this scheme. First, some parameters are used to carry over values from one time step to another. Second, for some elements the distinction between inputs and outputs cannot be made before the configuration is set up. This causality problem occurs in algebraic relations. However, the user will not have to worry about this problem since it is solved by the system during initialization.

In order to define a configuration the user draws up a block scheme of modules from the library. The connection points, or nodes, are numbered consecutively from one. Likewise, the modules in the configuration are given a consecutive order, which is reflected in the specification of input data.

The specification of input data can best be illustrated through an example. A simple configuration of three modules is shown in Fig. 4.3.

Module 1 is a comparator which has type number 1, module 2 is a multiplier which has type number 3 and module 3 is complex poles which has type number 42. The input data to be given is then:

- the number of modules (3)

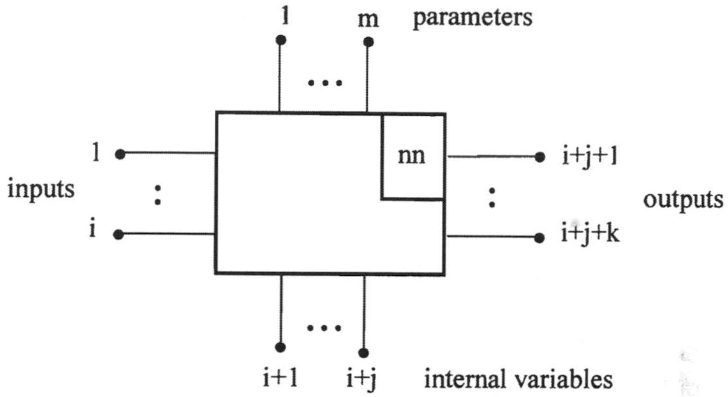

Figure 4.2: The general control module.

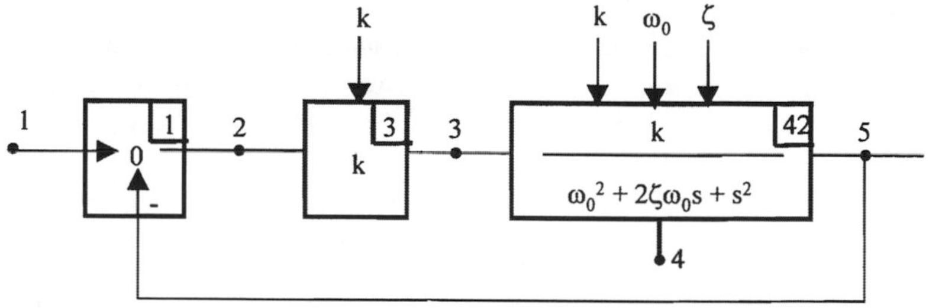

Figure 4.3: Configuration example.

- the number of nodes (5)
- a vector with the node numbers for each module (1,5,2;2,3;3,4,5)
- a vector with type and parameters for each module (1;3,k;42,k,ω_0,ζ)
- values for the initial states
- labels for the types of variables (1,0,0,0,0)

Label value 1 signifies input. The system will replace the zeros with twos for state variables or threes for algebraic variables. The vector of labels then becomes (1,3,3,2,2).

4.2.3 Software Structure

Overall Structure

In order to keep the interface clean and simple, the FE part is obliged to communicate with the control system through one single subroutine, which manages all activities for these modules. The three main activities are initialization, steady state computation and integration, as shown in the overall diagram in Fig. 4.4.

The software is further organized so as to concentrate the dependencies on the numerical methods and the module library. The purpose of using two numerical methods is for local error estimation. The numerical method dependencies are delimited to the subroutines for solution with backward Euler and Lobatto IIIC and for the generation of Newton matrices. The task for the library modules is essentially to perform the function evaluations of Eq. (4.61). Only the lower two levels in the diagram are affected by changes in the library.

Initialization

The task during initialization is mainly to label the state and algebraic type of variables, and for the algebraic part of Equation 4.61 to determine the variable to which the function value should be allocated. Each module must be prepared for this task which is managed from the initialization routine. Several iterations may be necessary in the case of algebraic loops or causality problems.

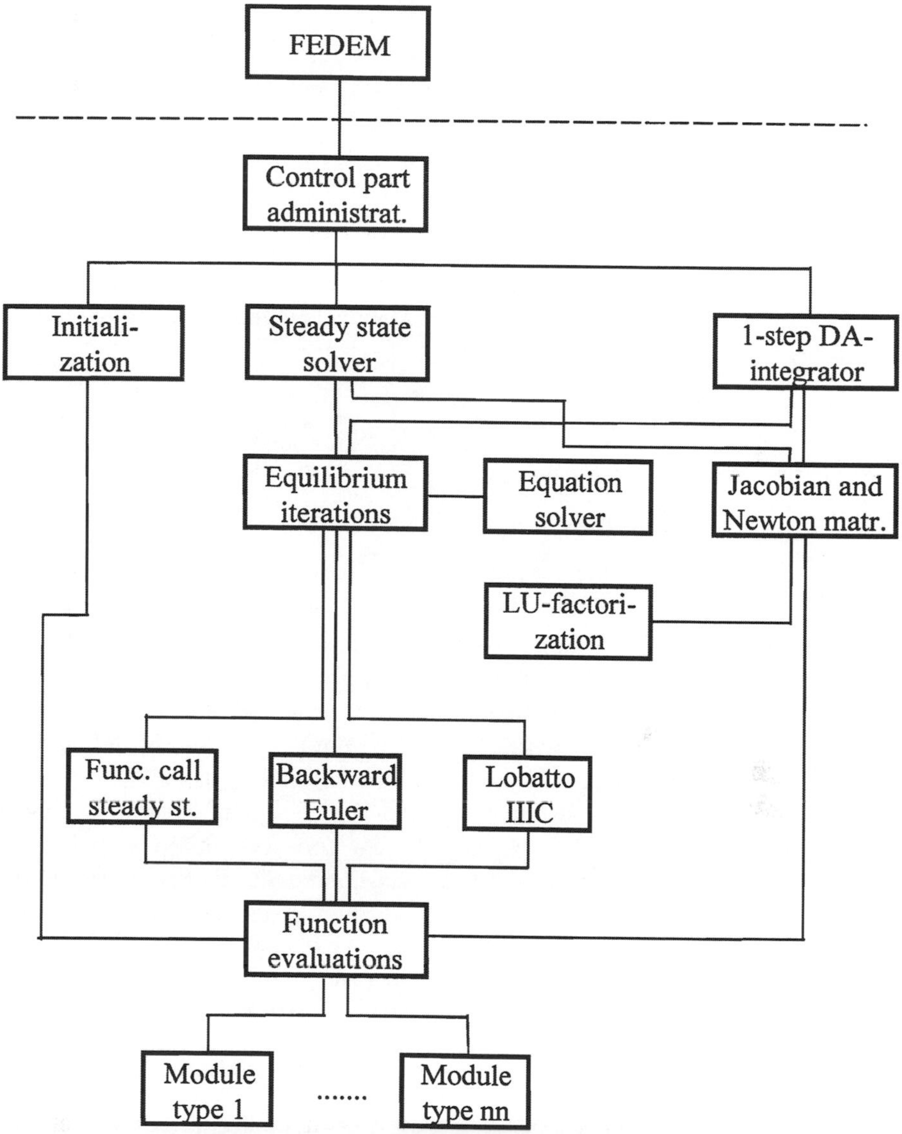

Figure 4.4: Overall software structure.

Steady State

With the different types of variables collected in one vector

$$\boldsymbol{y} = \begin{bmatrix} u & x & z \end{bmatrix}^T \tag{4.62}$$

the set of equations may be written:

$$\boldsymbol{F}(t, \boldsymbol{y}, \dot{\boldsymbol{y}}) = \begin{bmatrix} -\boldsymbol{u} + \boldsymbol{s}(t) \\ -\dot{\boldsymbol{x}} + \boldsymbol{f}(t, \boldsymbol{y}) \\ \boldsymbol{g}(t, \boldsymbol{y}) \end{bmatrix} = \begin{bmatrix} \mathbf{0} \\ \mathbf{0} \\ \mathbf{0} \end{bmatrix} \tag{4.63}$$

Steady state is found by setting the derivative to zero. Applying Newton-iteration to Eq. (4.63), we get the following system to solve at iteration number k:

$$\begin{bmatrix} \boldsymbol{I} & \mathbf{0} & \mathbf{0} \\ \mathbf{0} & -\boldsymbol{f}_x & -\boldsymbol{f}_z \\ \mathbf{0} & -\boldsymbol{g}_x & -\boldsymbol{g}_z \end{bmatrix} \begin{bmatrix} \Delta \boldsymbol{u}^k \\ \Delta \boldsymbol{x}^k \\ \Delta \boldsymbol{z}^k \end{bmatrix} = \begin{bmatrix} \mathbf{0} \\ \boldsymbol{f}(t, \boldsymbol{u}, \boldsymbol{x}^k, \boldsymbol{z}^k) \\ \boldsymbol{g}(t, \boldsymbol{u}, \boldsymbol{x}^k, \boldsymbol{z}^k) \end{bmatrix} \tag{4.64}$$

A standard equation solver is used for the solution of this vector equation.

Time Integration

For development of the numerical equations we take Eq. (4.63) as a starting point. Consequently, all of the input components can be treated as time functions, although some of them are determined by the FE part. None of the inputs will be changed in the control part. Due to the labeling, the variables in vector $\boldsymbol{y}$ need not be ordered according to type.

Since Newmark's β-method is of second order, it is natural to choose a method of second-order for the control part too. For this purpose we have chosen Lobatto IIIC which is an implicit, second-order Runge-Kutta method. Backward Euler, which is implicit too and first-order, is used for local error estimation. Local extrapolation is used, which means that the integration proceeds with the results from the higher order method.

The general m-level Implicit Runge-Kutta (IRK) method for solution of Eq. (4.63) can be written:

$$F(t + c_i h, Y_i, \dot{Y}_i) = 0 \tag{4.65}$$

$$Y_i = y_n + h \sum_{j=1}^{m} a_{ij} \dot{Y}_j \tag{4.66}$$

$$y_{n+1} = y_n + h \sum_{i=1}^{m} b_i \dot{Y}_i \tag{4.67}$$

Table 4.1: Butcher Tableau.

$$\begin{array}{c|c} \boldsymbol{c} & \boldsymbol{A} \\ \hline & \boldsymbol{b}^T \end{array}$$

Table 4.2: Backward Euler and Lobatto IIIC.

$$\begin{array}{c|c} 1 & 1 \\ \hline & 1 \end{array} \qquad \begin{array}{c|cc} 0 & 1/2 & -1/2 \\ 1 & 1/2 & 1/2 \\ \hline & 1/2 & 1/2 \end{array}$$

where h is the time step. In order to visualize a particular method the matrix $\boldsymbol{A}$ and vectors $\boldsymbol{b}$ and $\boldsymbol{c}$ formed by the a, b and c coefficients are usually put in a Butcher tableau as shown in Table 4.1. For backward Euler and Lobatto IIIC, we have the following Butcher tableaux given by Table 4.2. For an RK method satisfying

$$b_i = a_{mi}, \quad \text{for} \quad i = 1, \ldots, m \tag{4.68}$$

such as the two methods above, we get recording of the response by the equality:

$$y_{n+1} = Y_m \tag{4.69}$$

From Eqs. (4.68) and (4.69) an explicit expression may be derived for Y:

$$\dot{Y}_i = \frac{1}{h} \sum_{j=1}^{m} d_{ij} \left(Y_j - y_n\right), \quad \text{where} \quad \{d_{ij}\} = \boldsymbol{D} = \boldsymbol{A}^{-1} \tag{4.70}$$

where the d_{ij}-coefficients are the elements in the matrix $\boldsymbol{D} \equiv \boldsymbol{A}^{-1}$. Eqs. (4.65) to (4.70) can then be replaced by:

$$F\left(t + c_i h, Y_i, \frac{1}{h} \sum_{j=1}^{m} d_{ij} \left(Y_j - y_n\right)\right) = 0, \quad \text{for} \quad i = 1, \ldots, m \tag{4.71}$$

$$y_{n+1} = Y_m \tag{4.72}$$

For Backward Euler this gives:

$$F\left(t_{n+1}, \tilde{y}_{n+1}, \frac{1}{h}\left(\tilde{y}_{n+1} - y_n\right)\right) = 0 \tag{4.73}$$

and for Lobatto IIIC:

$$F\left(t_n, Y_1, \frac{1}{h}(Y_1 + Y_2 - 2y_n)\right) = 0 \tag{4.74}$$

$$F\left(t_{n+1}, Y_2, \frac{1}{h}(-Y_1 + Y_2)\right) = 0 \tag{4.75}$$

The local error estimate is then

$$l_{n+1} = \|y_{n+1} - \tilde{y}_{n+1}\| \tag{4.76}$$

In order to find the solution of the implicit numerical Eq. (4.73) and Eqs. (4.74) to (4.76) we use Newton iteration as for the steady-state computations. As the starting values for the iterations we take

$$\tilde{y}^0_{n+1} = y_n + \frac{h_{n+1}}{h_n}(y_n - y_{n-1}) \tag{4.77}$$

$$Y^0_{1,n+1} = y_n \tag{4.78}$$

$$Y^0_{2,n+1} = \tilde{y}_{n+1} \tag{4.79}$$

The Jacobian, which is needed in the Newton matrices, is generated numerically. This is done by excitation of each variable in turn and recording the response of the other ones.

4.2.4 Program Library

A number of general elements constitutes the library for the control part. These elements represent basic algebraic or differential operators, the time dependent sample-and-hold function, basic linear transfer functions, the most frequently used controllers and also algebraic elements with discontinuities in one of their derivatives such as the logical switch, limitation and dead zone.

In the linear transfer functions a linear system is also included

$$\dot{\boldsymbol{x}} = \boldsymbol{A}x + \boldsymbol{B}u, \quad \boldsymbol{y} = \boldsymbol{C}x + \boldsymbol{D}u \tag{4.80}$$

where $\boldsymbol{u}$, $\boldsymbol{x}$ and $\boldsymbol{y}$ are the input, state and output vectors respectively, and $\boldsymbol{A}$, $\boldsymbol{B}$, $\boldsymbol{C}$ and $\boldsymbol{D}$ are matrices with appropriate dimensions.

For the sample and hold element the user meets no restrictions with respect to relations between the sample period and the numerical time step. However, in order to maintain the order of the method, the system has to adjust the time step in order to hit the sample points.

The reason for this is that each sample represents a discontinuous change in the variable. The discontinuity is said to be of order q when it occurs in the qth derivative of one of the variables and the lower order derivatives are continuous. Gear and Østerby (1984) have shown that the accuracy of the result will drop below the order p of the method when $q \leq p$, unless we hit the discontinuity points with the discretization. In our case with $p = 2$, this means that actions must be taken for discontinuities in the first derivative of the right-hand side of Eq. (4.61).

For regular sampling the discontinuity points can easily be forseen. This is not so for other types of elements like the logical switch or the dead zone. In these cases the discontinuity point must be found by interpolation. For this purpose we use an interpolant which is optimal for Lobatto IIIC. For development of interpolants for RK-methods, see for instance Enright et al. (1986).

4.3 Simulation Results

4.3.1 Primary Simulation Variables

The primary simulation variables are the super node orientation and position, the super node translational and rotational velocity and acceleration and the current super element transformation matrix. Also variables from the control simulation, control variables, are primary variables in the simulation.

The super nodes' orientation and position are represented as a (3×4) transformation matrix, see Eq. (2.54):

$$\boldsymbol{t}_{SN} = \begin{bmatrix} R_{11} & R_{12} & R_{13} & P_x \\ R_{21} & R_{22} & R_{23} & P_y \\ R_{31} & R_{32} & R_{33} & P_z \end{bmatrix} = [\boldsymbol{e}_x \; \boldsymbol{e}_y \; \boldsymbol{e}_z \; \boldsymbol{P}] \tag{4.81}$$

where $\boldsymbol{e}_x$, $\boldsymbol{e}_y$ and $\boldsymbol{e}_z$ are the unit vectors in the x-, y- and z-directions, respectively, of the super node coordinate system and the vector $\boldsymbol{P}$ represents the position of super nodes, all referred to the global coordinate system. The three unit vectors represent the direction cosine matrix that uniquely defines the orientation in space. This orientation could also be expressed uniquely by three *Euler angles* alternatively to the overdetermined direction cosine representation. Regarding Euler angles see for instance Eq. (2.82).

The super node velocities are presented by two Cartesian vectors, one representing the translational velocity and one the rotational velocity relative to the global coordinate system as shown below

$$\begin{bmatrix} v_{tx} & v_{rx} \\ v_{ty} & v_{ry} \\ v_{tz} & v_{rz} \end{bmatrix} = [\boldsymbol{v}_t \; \boldsymbol{v}_r] \tag{4.82}$$

where

$\boldsymbol{v}_t$: translational velocity vector

$\boldsymbol{v}_r$: rotational velocity vector (angular velocity)

The super node accelerations are presented in a very similar way as two Cartesian vectors relative to the global coordinate system as shown below:

$$\begin{bmatrix} a_{tx} & a_{rx} \\ a_{ty} & a_{ry} \\ a_{tz} & a_{rz} \end{bmatrix} = [\boldsymbol{a}_t \; \boldsymbol{a}_r] \tag{4.83}$$

where

$\boldsymbol{a}_t$: translational acceleration vector

$\boldsymbol{a}_r$: rotational acceleration vector (angular acceleration)

The magnitudes for the modal super node displacement, velocity and acceleration, representing super element modes of vibration, are presented in a similar way as ordinary super nodes, except there are a variable number of DOFs per super node. The variables for a modal super node may be written:

$$\begin{bmatrix} y_1 & \dot{y}_1 & \ddot{y}_1 \\ y_2 & \dot{y}_2 & \ddot{y}_2 \\ \cdot & \cdot & \cdot \\ y_n & \dot{y}_n & \ddot{y}_n \end{bmatrix} = [\boldsymbol{y} \; \dot{\boldsymbol{y}} \; \ddot{\boldsymbol{y}}] \tag{4.84}$$

where

$\boldsymbol{y}$: modal super node displacement magnitude

$\dot{\boldsymbol{y}}$: modal super node velocity magnitude

$\ddot{\boldsymbol{y}}$: modal super node acceleration magnitude

The super element position and orientation are represented as a (3×4) transformation matrix, see Eqs. (2.54) and (4.81):

$$\boldsymbol{t}_{SE} = \begin{bmatrix} R_{11} & R_{12} & R_{13} & o_x \\ R_{21} & R_{22} & R_{23} & o_y \\ R_{31} & R_{32} & R_{33} & o_z \end{bmatrix} = [\boldsymbol{e}_x \; \boldsymbol{e}_y \; \boldsymbol{e}_z \; \boldsymbol{o}] \tag{4.85}$$

where the unit vectors $\boldsymbol{e}_x$, $\boldsymbol{e}_y$ and $\boldsymbol{e}_z$, represent the orientation of the super element coordinate system, while $\boldsymbol{o}$ represents the position for the origin of the super element coordinate system.

As mentioned above control variables are also primary simulation variables. The control variables are defined as input, state and algebraic variables, that means for example the variables in nodes 1 to 5 in the control configuration example in Fig. 4.3.

4.3.2 Derived Simulation Variables

Some of the simulation variables are derived from the primary variables. This is the case for

- Joint variables
- Spring variables
- Damper variables
- Forces and torques

Definition of joint variables for different joints are derived from the primary variables of the super nodes involved in the actual joint, refer to Section 3.4 for joint variable definition. Results may be required for joint variable position, velocity and acceleration. Each joint may have one or more variables, for example the revolute joint has 1 variable while the ball joint has 3 variables.

A spring element, being a uniaxial or a joint spring, will have three variables: the spring length/angle, the spring deflection from the stress-free position and the spring force/torque. For a uniaxial spring the spring length is calculated from the position of the actual two end nodes. For joint springs, however, the spring length or angle will refer to a joint variable of the actual joint. Subtracting the specified stress-free length/angle of the spring for the actual position from the calculated length/angle of the spring will give the spring deflection. In the general nonlinear case the spring force/torque is found by integrating the spring stiffness over the spring deflection.

A damper element, being a uniaxial or joint damper, will have three variables: the damper length/angle, the damper relative velocity and the damper force/torque. For uniaxial dampers the damper length is calculated from the position of the actual two end nodes. For joint dampers, however, the damper length or angle will refer to a joint variable of the actual joint. Damper velocity for uniaxial dampers is calculated from the end node velocities. If we denote the global coordinates of the two end points of the damper

$$\boldsymbol{P}_1 = \begin{bmatrix} x_1 \\ y_1 \\ z_1 \end{bmatrix} \quad \textit{and } \boldsymbol{P}_2 = \begin{bmatrix} x_2 \\ y_2 \\ z_3 \end{bmatrix} \tag{4.86}$$

then the damper unit direction vector is calculated from

$$\boldsymbol{e}_{12} = \frac{\boldsymbol{P}_2 - \boldsymbol{P}_1}{\|\boldsymbol{P}_2 - \boldsymbol{P}_1\|} \tag{4.87}$$

The relative velocity for these two points may be calculated from

$$\boldsymbol{v}_{12} = \begin{bmatrix} \dot{x}_1 \\ \dot{y}_1 \\ \dot{z}_1 \end{bmatrix} - \begin{bmatrix} \dot{x}_2 \\ \dot{y}_2 \\ \dot{z}_3 \end{bmatrix} \tag{4.88}$$

The damper velocity is then calculated from the scalar product

$$\boldsymbol{v}_D = \boldsymbol{e}_{12} \cdot \boldsymbol{v}_{12} \tag{4.89}$$

For joint dampers, however, the damper velocity refers directly to a joint variable velocity. The damper force/torque is found by multiplying the actual damping coefficient by the actual damper velocity.

For calculation of super element node forces and torques, refer to the stiffness relation of Eq. (3.42) Section 3.3. These forces are transformed to global direction and added into the system vector. This system vector is then used to present the nodal forces and torques relative a global coordinate system. Joint forces and torques are found on the corresponding master or slave DOFs of the actual joint. See the calculation of stiffness forces in Section 4.1.

4.3.3 Substructure Retracking

The updated super element formulation developed in Section 3.3, shows the derivation of the super element deformation vector for an external node, Eq. (3.40). The modal coordinates $\boldsymbol{y}_i$ are primary integration variables, and the super element deformation vector for all external nodes and modal coordinates are shown in Eq. (3.41). The inverse CMS transformation gives

$$\begin{aligned} \boldsymbol{v}_i &= \boldsymbol{v}_i^i + \boldsymbol{v}_i^e \\ &= \boldsymbol{\Phi} \cdot \boldsymbol{y} + \boldsymbol{B}\boldsymbol{v}_e \end{aligned} \tag{4.90}$$

and the expanded substructure displacement vector may be written

$$\boldsymbol{v} = \begin{bmatrix} \boldsymbol{v}_e \\ \boldsymbol{v}_i \end{bmatrix} \tag{4.91}$$

see Eqs. (3.9) and (3.19).

Nodal displacements for substructure primary elements may now be extracted from the substructure displacement vector.

$$\boldsymbol{u}_j = \boldsymbol{b}_j \boldsymbol{v} \tag{4.92}$$

where

$\boldsymbol{u}_j$: represents the displacement vector of primary element j in the substructure.

$\boldsymbol{b}_j$: the incidence matrix that represents the topology for primary element j in the substructure.

$\boldsymbol{v}$: represents the substructure displacement vector

4.3.4 Finite Element Stress Analysis

The displacements within a finite element are expressed by the nodal degrees of freedom $\boldsymbol{u}$ and a set of interpolation polynoms N_i represented by the vector $\boldsymbol{N}$, for instance for a membrane finite element.

$$u(x, y) = \boldsymbol{N}\boldsymbol{u} \tag{4.93}$$

where

$$\boldsymbol{N} = [N_1(x, y), N_2(x, y), ..., N_n(x, y)] \tag{4.94}$$

and

$$\boldsymbol{u} = [u_1, u_2, ..., u_n] \tag{4.95}$$

The shape functions are chosen so that $N_i(x, y) = 1$ for DOF u_i and zero for DOFs in all the other nodal points within an element. It may be shown that the strains $\boldsymbol{\varepsilon}$ within a finite element may be calculated from:

$$\boldsymbol{\varepsilon} = f(\boldsymbol{N}\boldsymbol{u}) = \bar{\boldsymbol{B}}\boldsymbol{u} \tag{4.96}$$

where

$\boldsymbol{\varepsilon}$: the element strain vector

$\bar{\boldsymbol{B}}$: strain function matrix

and the stress according to Hooke's Law:

$$\boldsymbol{\sigma} = \boldsymbol{E}\boldsymbol{\varepsilon} \tag{4.97}$$

where

$\boldsymbol{\sigma}$: the element stress vector

$\boldsymbol{E}$: elasticity matrix

For details, refer to textbooks on the finite element method.

4.3.5 Virtual Strain Gauges

The modeling of virtual strain gauges is a new feature in the simulation code that has been recently developed. This means that time series of the strains and stresses can be generated from a simulation run. In many cases this can replace physical testing in the laboratory as it provides input to fatigue analysis of the assembled product from lifelike operations. This gives a new dimension to virtual testing, a term that is becoming used more and more often now.

4.3.6 Eigenvalues Results

Eigenvalue analysis at system level evaluates a number of the lowest eigenvalues for the structure. The result from the eigenvalue analysis is the natural frequencies and the corresponding eigenvectors representing the different modes of vibration. These results are presented at specified positions during a simulation session.

4.4 Energy Calculations

The energy calculations are divided into the following terms (the text in this section is adopted from the Fedem Theory Manual (1999)):

U_ε Strain energy, computed for all links, springs and system (system is sum of all).

U_k Kinetic energy, computed for all links, discrete masses, rotational inertias and system.

U_p Potential energy, computed for all links, discrete masses, and system.

U_i Input energy, computed for all external forces, springs (contribution from nonconstant stress-free length), and system.

U_d Energy loss, computed for all links (structural damping), axial damper, joint dampers, joint friction, and system.

U_{sum} Energy check-sum, computed for the system.

All these terms can be plotted as functions of time (or other variables).

4.4.1 Strain Energy

The total strain energy for the system is computed as a sum of all the element strain energies and the spring strain energies.

Link Strain Energy

The strain energy for one link (super-element) with linearly elastic material is given by

$$U_\varepsilon = \frac{1}{2}\boldsymbol{v}_d^T \boldsymbol{K} \boldsymbol{v}_d \tag{4.98}$$

where $\boldsymbol{v}_d$ is the deformational displacements of the element, and $\boldsymbol{K}$ is the element stiffness matrix. The strain energy is thus computed on a total form at each converged time-step; i.e. no storage of previous strain energy is necessary.

Spring Strain Energy

Even the nonlinear springs have a hyper-elastic behavior, and the strain energy can thus be computed in a total form at each converged time-step.

$$\begin{aligned} &\text{Linear springs :} \qquad && U_\varepsilon = \frac{1}{2}k_s u^2 \\ &\text{Nonlinear springs:} \qquad && U_\varepsilon = \int_0^u f_s(\hat{u})\, d\hat{u} \end{aligned} \tag{4.99}$$

Since the expression for the hyper-elastic nonlinear spring energy also covers the linear spring element, this is thus used for all spring elements; linear and nonlinear, axial and joint springs. There is no need for an incremental calculation for the spring strain energy as long as only hyper-elastic nonlinear springs are included in the model.

4.4.2 Kinetic Energy

System kinetic energy consists of the contributions from all the links, all the lumped masses, and all the discrete rotational inertias.

Link Kinetic Energy

Calculation of the link kinetic energy is done based on the super-element mass matrix, $\boldsymbol{M}$, and the super-element velocity vector; $\dot{\boldsymbol{v}}$, as:

$$U_k = \frac{1}{2}\dot{\boldsymbol{v}}^T \boldsymbol{M}\dot{\boldsymbol{v}} \tag{4.100}$$

i.e. also computed on a total form at each time.

Kinetic Energy from Discrete Masses and Inertias

Discrete masses and rotational inertias contribute to the kinetic energy as:

$$U_k = \frac{1}{2}m\dot{\boldsymbol{u}}^T\dot{\boldsymbol{u}} + \frac{1}{2}\boldsymbol{\omega}^T\boldsymbol{I}_\omega\boldsymbol{\omega} \tag{4.101}$$

where m is the lumped mass and $\boldsymbol{I}_\omega$ is the rotational inertia tensor for the lumped mass.

4.4.3 Potential Energy

As with the kinetic energy, the system potential energy is the sum of all the contributions from the links and the discrete masses. However, the discrete rotational inertias do not contribute to the potential energy.

Link Potential Energy

Computation of the potential energy should reflect the changes in potential energy rather than the potential energy relative to the coordinate system used for modeling. This prevents hiding other energy contributions by choosing a coordinate system which gives large potential energies. This is achieved by subtracting the potential energy of the masses in their initial position C_0, from the potential energy at present position C_n i.e. the initial potential energy of all masses, super-element masses and lumped masses, are zero.

$$U_p = m\boldsymbol{g}^T\left(\boldsymbol{x}_{C_n} - \boldsymbol{x}_{C_0}\right) \tag{4.102}$$

Calculations of the potential energy of the links use the displacements of the link centroid. This calculation neglects the relative displacement of the centroid due to the internal deformations of the super-elements, however this is justified within the assumption of small deformations. The calculation of the potential energy is performed in a total fashion without storing energies from previous steps.

4.4.4 Input Energy

Input energy to the system consists of contributions from external forces and springs that have nonconstant stress-free length.

Input Energy from External Forces

Since the external forces are nonconservative (can have a time variation and corotational behavior) the input energy from the external forces must be computed in an incremental manner from one time-step to the next:

$$U_i = \int_0^t F\left(\hat{t}\right) \dot{u} d\hat{t} = \sum_{k=1}^{n} \overline{F}_k^T \Delta u_k \quad \text{where} \quad \overline{F}_k = \frac{1}{2}\left(F_{k-1} + F_k\right) \tag{4.103}$$

Subscript k above is the time-step index. The summation expression above represents a trapezoidal integration scheme.

Input Energy from Springs

The stress-free length of the springs (both axial springs and joint springs) are subject to possible change in stress-free length. When the spring is stressed this represents an input energy contribution. This input energy must be computed incrementally for all the springs

$$U_i = \int_0^t f_s(\hat{t}) l_0 \, d\hat{t} = \sum_{k=1}^{n} \overline{f}_{s_k} \Delta l_0 \quad \text{where} \quad \overline{f}_{s_k} = \frac{1}{2}(f_{s_{k-1}} + f_{s_k}) \tag{4.104}$$

4.4.5 Energy Loss

The total energy loss consists of the contributions from structural damping of the links, discrete dampers (both axial dampers and joint dampers), and energy loss from joint friction.

Energy Loss from Structural Damping in Links

The structural damping of the links is composed of mass and stiffness proportional damping i.e. $\alpha_1 \boldsymbol{M} + \alpha_2 \boldsymbol{K}$, see Eq. (4.37), which is used in the following damping energy expression for a link

$$U_d = \int_0^t \dot{v}^T C \dot{v} d\hat{t} = \sum_{k=1}^{n} h\alpha_1 \dot{v}_k^T M v_k + \sum_{k=1}^{n} h\alpha_2 \dot{v}_{d_k}^T K v_{d_k} \tag{4.105}$$

where $\dot{v}_k^T = \frac{1}{2}\left(v_{k-1} + v_k\right)$, $\dot{v}_{d_k} = \frac{1}{h}\left(v_{d_k} - v_{d_{k-1}}\right)$ and h represents the time-step size.

Note that the deformational velocities, $\dot{v}_d$, have been used to compute the stiffness proportional damping energy. This is important in order to avoid fictitious damping energies from rigid body velocities when using fairly large time-steps.

Energy Loss from Discrete Dampers

Energy loss from discrete dampers (axial dampers and joint dampers) is computed as

$$U_d = \int_0^t f_d \dot{u}\, dt = \sum_{k=1}^{n} \overline{f}_{d_k} \Delta u \tag{4.106}$$

where $\overline{f}_{d_k} = \frac{1}{2}\left(f_{d_{k-1}} + f_{d_k}\right)$

Energy Loss from Friction

Friction energy loss is computationally very analogous to the damping loss

$$U_d = \int_0^t f_f\, dt = \sum_{k=1}^{n} \overline{f}_{f_k} \Delta u_k \tag{4.107}$$

where $\overline{f}_{f_k} = \frac{1}{2}\left(f_{f_{k-1}} + f_{f_k}\right)$

4.4.6 Energy Check-sum

An energy check-sum can be computed for the system in order to verify that the system energy is preserved:

$$U_{sum} = U_{p_{system}} + U_{k\,system} + U_{\varepsilon\,system} + U_{d\,system} - U_{i\,system} \tag{4.108}$$

The check-sum should remain constant during time domain dynamic analysis.

Chapter 5

Dimensioning of Mechanisms

5.1 Interactive Modeling and Simulation

5.1.1 Introduction

The modeling of mechanisms is often one of the most challenging tasks that a mechanical engineer could be involved in. Modern machine design usually involves a number of disciplines including static analysis, structural dynamics, vibrations, motion analysis, control engineering, electronics, styling, economics, etc., but these are also fields that fall into other engineering disciplines. This means that civil engineers, control engineers, electrical engineers, designers and economists have one or more machine design subjects as one of their main tasks. Kinematics and mechanism design is, however, one of the most typical domains for a mechanical engineer, but often many or all the above-mentioned disciplines are also involved in making a mechanism an integrated part of a product.

To start the design of a mechanism the mechanical engineer has to select a mechanism topology that can give the required motion. Synthesis programs may be used to calculate the overall mechanism dimensions such as the position and direction of joints for the mechanism design position to give the required constrained motion. In other situations, the mechanism design position is developed in a CAD system that is often based on simple kinematics analysis. In order to optimize or verify high precision mechanisms with respect to link flexibility, weight and control, a functional mechanism model may be developed in the multidisciplinary simulation code FEDEM. Multidisciplinary simulation could be used for verifying a design, however, such a tool is more often used as a design tool starting with a feasibility study and continuously refining the model to an optimized and finished design. This chapter will focus on dimensioning of high performance and multidisciplinary mechanisms where the theory presented in this book is needed to optimize and verify the design.

The FEDEM computer code is based on the theory presented in this book. For the commercialization of the program an interactive graphic interface has been developed to simplify the user input to the simulation algorithms. As described in earlier chapters of the book the simulation is based on a Finite Element (FE) approach and the mechanism links are modeled individually as FE meshes in a standard FE pre-processor and imported into the FEDEM code during mechanism modeling. Mechanism simulation in FEDEM is menu-based and organized in the three tasks: modeling, simulation and post-processing.

The modeling task has menus for importing FE links, positioning links, positioning joints, attaching joints to links, entering springs and dampers, positioning forces and torques, adding function input, etc. A separate window may be entered for control modeling from a library of control elements with input and output terminals that are connected in a control scheme. Control input elements are used to connect the input of the control to the mechanism model through position, velocity and acceleration measurements and control output elements for outputting force and torque actuators to the mechanism model, for instance for closing control loops.

When the mechanism model is completed the simulation task is entered. In this task parameters are entered for controling the dynamic simulation, possible stress retracking and modal analysis at system level. Most of the time integration parameters have default values, however, the integration time and the time step must be entered for the actual simulation. The mechanism model is animated during simulation. Prior to stress retracking, time positions from the completed time integration must be specified for stress calculations. Modal animation may also be specified in the simulation task.

When the simulation task computations have been completed, the post-processing task should be entered to inspect the results. The curve-plotting module may be used for showing the response of a simulation variable as a function of time or as function of another simulation variable; typically for displaying a coupler curve. If stress retracking has been carried out during the simulation task, the high performance animation module may be started to display mechanism animation including stress contour plots and deformations. The simulation post-processor may also display any modal animation results generated in the simulation task.

5.1.2 Multidisciplinary Modeling

A design position for the mechanism is often available prior to the multidisciplinary modeling in FEDEM, for instance from a CAD system or from a synthesis program as mentioned earlier. However, in many cases the modeling starts from a preliminary sketch where positions of joints and preliminary link shapes are available and possible control laws are written down. The modeling starts in a FE pre-processor where a FE mesh is generated and FE element physical and material properties are selected. FE models for the different links of the mechanism are exported in separate files as input to the mechanism modeler. All connection points for entities in the mechanism model are restricted to positions of the FE nodes generated by the FE pre-processor and this has to be taken into account when the FE mesh is generated. The FE nodes selected as connection points for mechanism entities are denoted as triads or super nodes in the mechanism model.

With files containing FE models available for all the mechanism links, the modeling could start in the mechanism modeler. The link FE-models are imported into the graphic window one after the other and positioned and oriented relative to a global coordinate system. After positioning the links the actual FE nodes of the imported meshes must be in positions where joints, springs, dampers, forces, torques, additional inertias, control sensors and coupler points should be attached.

Two types of joints are available, point-to-point and point-to-line joints. A joint may connect two and only two links, however, one of the links may be the ground link. For point-to-point joints such as revolute, ball and rigid joints, a FE node from each of the two involved links must be in the same position, the position where the joint centre is positioned. After positioning the joint entity to the position of these co-incident FE nodes the joint is, if required, rotated to the direction of the joint axis. The joint can then be attached to the actual links. For point-to-line joints such as prismatic and cylindrical joints one node (the slave) on the first link is connected to a line of nodes (the masters) on the other link. Attaching these nodes the relative motion between the links is constrained to sliding along this line. For the cylindrical joint rotation about the line defined by the master nodes is also possible. Mechanical transmission models are based on the joint modeling, for instance two revolute joints on one link, the gear housing, and may be coupled with a gear ratio. A rack and pinion transmission is based on a revolute joint and a prismatic joint. A screw joint is modeled by specifying a screw ratio for a cylindrical joint. A free joint is also a point-to-point joint, however, the nodes of the two links do not have to be co-incident. The free joint only has spring constraints.

For this FE-based formulation with flexible links there is no limitation of only one joint between two links as for rigid body mechanism formulations. That means you may have as many parallel joints between links as you like and you can study the effects from production tolerances by specifying deviations for joint axis directions for instance for parallel revolute joints. Another example where tolerance deviations may be studied is the planar four bar mechanism. If the direction of the joint axis is modeled with a small deviation from parallel to the other joint axis the simulation will run. However, the links will have deformations and stresses from the twisting of the links during rotation of the input links.

Springs and dampers are other basic mechanism entities. Springs and dampers within joints are modeled as part of the joint entity. Axial springs and dampers are usually connected between FE nodes from different links. All springs and dampers may be linear or nonlinear. A linear spring or damper is specified by entering the spring or damper constant. However, for nonlinear springs and dampers a separate function is specified for describing the actual nonlinear spring or damper characteristic. A wide range of nonlinear effects may be modeled in this way. A spring may also be used as a drive element (motor) for a mechanism by specifying a function for the spring stress free length as a function of integration time or as a function of another simulation variable. Springs represent very powerful tools for mechanism modeling.

The finite element models imported from the FE pre-processor introduce mass and inertia into the mechanism model. Nevertheless, the designer often needs to introduce extra point masses and inertias to the links. The extra mass or inertia could for instance be used to compensate for limitation in the FE model by adding extra mass in a joint or by introducing flywheel effects on motor shafts or for introducing the masses and inertias of objects to be handled by the mechanism. Mass and inertia are entered through triad entities. A gravitation vector for the system may be defined with direction and magnitude.

It is mentioned above that spring elements may be used as mechanism drive elements. Forces and torques are also drive elements that are modeled as vectors acting on the links. A FE node is identified where the force or the torque will be positioned. Two points are used to specify the direction of an input force or torque vector. Coordinates for these points are given individually relative to the global coordinate system (ground link) or relative to the updated coordinate system for any other link and this is a very flexible and powerful modeling tool. The magnitude for the force or torque may be constant or specified by a function definition. The force or torque magnitude may also be the output from a control system.

A closed loop control system may be composed from a library of control modules such as basic control modules, time dependent modules, piecewise continuous modules and compensatory modules. For the control system to have measurements from the mechanism, sensor elements must be defined in the mechanism model to measure positions, velocities or accelerations of selected simulation variables. Control loops are closed by feeding the output from the control system as magnitudes for forces or torques as mentioned above.

5.1.3 Setting of Simulation Parameters and Options

A model constructed by the interactive modeling tools described above is converted to a simulation model based on the FE-based formulation presented in this book. The primary goal for this simulation model is dynamic simulation using time integration. In general, time integration of mechanism motion is a very nonlinear problem and a Newmark integration scheme for nonlinear problems is adopted. An incremental integration formulation is used, that is the integration is advanced one time increment at a time, and for each time increment iterations are used to assure equilibrium at the end of the increment. The simulation task has three modules available, the dynamic simulation, the stress retracking and the modal analysis. A dynamic simulation run is required prior to a stress retracking run or a modal analysis run.

The simulation usually starts from integration time zero and is advanced at one increment at a time until the end of the simulation time set by the user is reached. The user also sets the increment for the time integration. All other settings and options for the time integration have default values. However, these settings and options should also be checked and corrected if necessary. Some of the optional settings and options that are most often adjusted are:

- Eigenvalue analysis
- Initial quasistatic equilibrium iteration and corresponding tolerance setting
- Time integration tolerance
- Integration parameter for Newmark α-damping
- Constants for proportional damping
- Maximum number of iterations in each time increment

- Variable integration time increment including lower and upper bounds for the increment
- Correction for stress stiffening

When the simulation is started the mechanism model in the graphic window is animated as rigid bodies simultaneously as the integration proceeds and an indicator is also updated to give the fraction of the simulation completed at any time. A simulation may be prolonged from the end simulation time using a restart option.

The module available for the retracking of deformations and stresses for the links or super elements involved in the dynamic simulation has a default setting for stress analysis for each tenth time step. The user may edit the time settings for stress analysis, but the settings must be a subset of the time increments from the dynamic simulation.

For time increments where eigenvalue analysis has been specified during dynamic simulation the modal analysis module may be executed for generating deformed geometry for eigenvector animation. A number of mechanism geometry sets may be generated for each eigenvector, in each specified position for eigenvalue analysis during dynamic simulation. Scaling factors for eigenvalue components will be for instance 10 evenly distributed values within one period of a sine curve and that is the basis for generation of the deformed geometry sets.

5.1.4 Visualization of Simulation Results

During simulation the mechanism positions will be continuously updated and displayed as animation of rigid bodies in the modeler graphic window. This rigid body animation will keep the user informed about how the simulation is proceeding and if the mechanism is behaving as expected. Entering the post-processing task this rigid-body animation may be repeated and inspected in more detail. After completing a simulation the curve-plotting module is used to display time responses of the mechanism system variables such positions, velocities, accelerations for triads and joint variables. Forces and torques in joints (constraining forces), springs and dampers may also be displayed as response curves, that is the simulation variable as a function of time. A curve for a simulation variable may also be plotted as a function of another simulation variable, typically the y-position of a triad can be plotted as a function of the x-position of the same triad, that is for displaying the coupler curve for that triad.

If the simulation task was also used to execute the stress analysis module the results could be inspected using the high performance visualization module. This module can also be used for rigid-body mechanism animation, however, this visualization tool is mainly used for showing mechanism animation including deformed bodies and/or stress contour plots. Using scaling factor 1 for the deformation of the links, all links will fit exactly together in the joints. However, using no scaling or a scaling factor different from 1 for the deformation of the link super nodes that should be coincident in joints may have deviations, but this is no error and is as expected. The sizes for these deviations in joints during visualization is to a large extent dependent on the position of the link corotated coordinate systems and of the size of the scaling factor itself.

For eigenvectors where geometry sets for animation were generated in the simulation task, the high performance visualization module may be executed in the post-processing task.

5.2 Case Study with FEDEM

5.2.1 Case Description

In this Section a four-bar path following mechanism is selected to demonstrate the design process for a mechanism using the multidisciplinary simulation approach presented in this book. The mechanism dimensions are selected for generating straight line motion, see Fig. 5.1. The system has four links including a ground link: an input link between points A and C, the coupler link between points C, B and D, the output link between points B and E and finely the ground link between point A and E. There are revolute joints in A, C, B and E and a coupler point in D. An electric motor will be used to give input motion for the revolute joint in A and a simple control system is used to control the rotational speed of this motor. The mechanism should have low weight and be stiff for deformation that is out of the plane for the motion. For all links the same cross-section dimensions are proposed, 5 mm thick and 18 mm wide, and aluminium is selected as link material. The links will be mounted with the width perpendicular to the plane of motion that is with quite high flexibility in the plane of motion. The position of the mechanism in Fig. 5.1 is selected as the design position for the modeling in the FEDEM simulation program. The input link should have a rotational speed of one revolution per second and the aim is to reach this rotational speed in 0.5 seconds.

The rotor mass-moment of inertia for the motor is chosen to be $0.15\ kgm^2$. The main goals for the simulation are to tune the control amplification and the motor flywheel inertia such that:

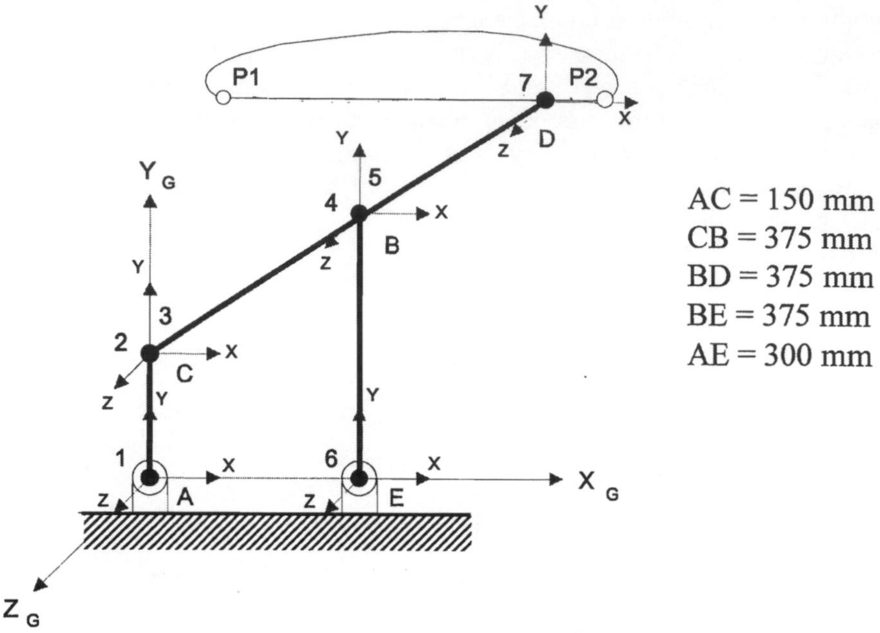

Figure 5.1: Path following mechanism.

- The input rotational velocity is between 5 and 8 rad/sec and the torque on the motor shaft should be as low as possible.
- The acceleration in the x-direction for point D must be below 40 m/sec^2.

As a result of the simulation:

- The torque on the motor shaft in the simulation is used to select the size of the motor.
- Position, velocity and acceleration in x- and y-directions are plotted as a function of time for point D, the coupler point.
- The rotational velocity and the torque for the motor shaft are plotted as a function of time.
- The coupler curve for point D is plotted.
- The stresses in the links are checked when steady-state periodic motion is achieved. What can be said about the fatigue life for this mechanism and where it is most likely that a fatigue crack will start?

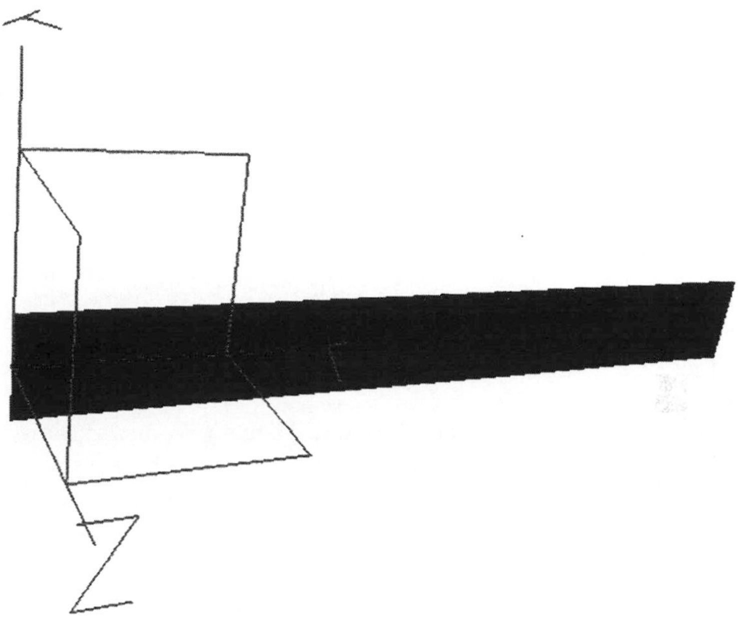

Figure 5.2: FE mesh for the mechanism link AC.

The first modeling step will now be to make the FE flexible link input models for the simulation. The FE-models for a FEDEM simulation could be made in for instance I-DEAS, NASTRAN or other FE-modelers that are interfaced to the FEDEM program.

5.2.2 FE-modeling

All the three flexible links are supposed to have the same cross sectional dimensions of 5 by 18 millimetres and be made from aluminium. A rectangular shell element in this case will be a good choice for the FE model. In order to model the link AC a plain surface of 18 by 150 millimetres is needed and a FE node is needed in the middle of both the end edges of this surface for connection to joints in the simulation model. Meshing with quadrilateral shell elements of element length 9 millimetres is selected that will produce two rows of FE elements along the link, see Fig. 5.2. The output link (BE) is modeled in a very similar way as the input link, however, the length is 375 millimetres. Also this link is meshed from two rows of quadrilateral shell elements along the link. The coupler link (CBD) is also modeled in the same way as the other links, however, in order to get the mechanism parts to fit exactly together as shown in Fig. 5.1 the mesh must include a FE node exactly in the middle of the 750 millimetre long link.

The dimensions of the FE mesh of the links are converted from millimeters to meters and are then exported in separate files in a file format that can be read by the FEDEM simulation code.

5.2.3 Assembling the Mechanism Model

Having the FE meshes for the mechanism links prepared and exported in the FEDEM input format, the FEDEM program is started to execute the modeling task. Gravitation of 9.81 m/sec^2 is specified in the negative global y-direction.

The usual approach is to import the FE meshes one at a time and use the "Smart move" modeling function in FEDEM to position the links in their actual positions for the mechanism design position. For our case we could start with the input link. When the import function has been executed for this link the FE mesh will be displayed in the position where the local link coordinate system is coincident with the global mechanism coordinate system. The positioning of the input link could be as follows, refer to Fig. 5.1 above: The mesh is first translated so that the middle FE node of the links left edge is positioned in the origin of the global coordinate system. The link is then rotated about the middle left node to a vertical position and at last rotated about a vertical axis through the global origin so the mesh is parallel to the global yz-plane.

Next, the output link (BE) mesh could be imported. The middle left node of this mesh is translated to position (0.3, 0, 0). The mesh is then rotated about this position to a vertical position and then rotated to be parallel to the global yz-plane. Finally the coupler link (CBD) is positioned. The link is imported and the middle left node of this mesh is translated to the position of the top of the input link, position (0, 0.15, 0). The mesh is then rotated about this point to be parallel to the xz-plane. A horizontal line along the mesh through the middle left node is then rotated with the mesh about the same node to pass through the top of the output link, position (0.3, 0.375, 0). All links are then in the proposed design position.

Joint entities, in our case revolute joints, are positioned and attached to the actual links in A, C, B and E. The joints are positioned in FE nodes in the xy-plane and the joint axes are made parallel to the global z-axis. The

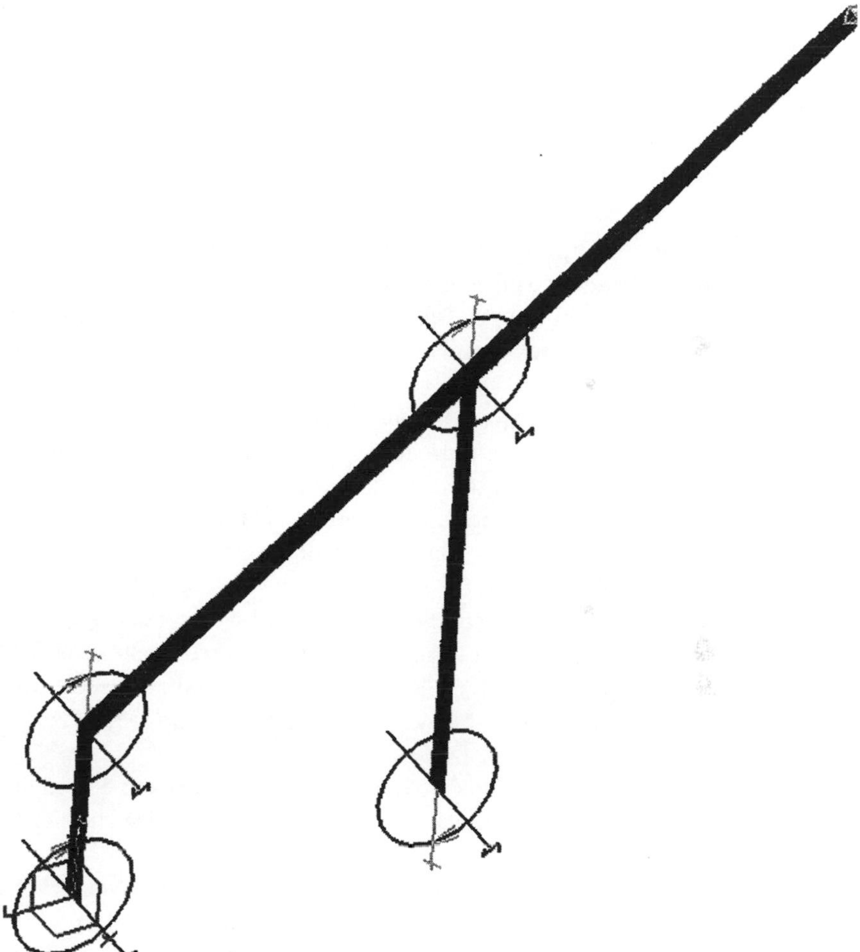

Figure 5.3: Joint symbols.

symbol for a revolute joint is a circle and the joint variable is marked by an arrow, see Fig. 5.3. Joints are attached at points A and E to the input and ground link and to the output and ground link, respectively. Joints are attached at points C and B to the input and coupler link and to the output and coupler link, respectively.

Inertia of 0.15 kgm^2 about the z-axis is entered on the triad or super node on the input link where joint A is positioned. This inertia represents the mass-moment of inertia for the rotor of the electric motor. This inertia could be increased to model a flywheel on the motor shaft. In order to have degrees of freedom in the simulation for the coupler point, point D, a triad is positioned in this point.

The torque from the input motor is modeled by a torque entity positioned on the same triad as the inertia above on the input link. This torque will have a direction that is opposite to the global z-axis. The magnitude for this torque is generated by a control system.

The control system should measure the rotational velocity of the revolute joint at point A and compare it with a prescribed time function for this velocity. The control system must evaluate this deviation and generate a torque to reduce this deviation. A sensor entity is modeled in the revolute joint at point A to measure the joint variable velocity. The reference for the rotational speed of the motor should be one revolution per second and the reference speed should increase from zero to one revolution per second in 0.5 seconds. This reference function for the control system is modeled as a limited ramp function entity having a slope of 12.56 rad/sec^2 in the interval between 0.0 and 0.5 seconds and constant of 6.28 rad/sec after 0.5 seconds.

A control system with two control input elements are modeled, the reference function and the rotational velocity measured by the modeled sensor entity. The function value and the measured velocity are compared in a control comparator module and the deviation is used to generate the motor torque. In this case the deviation between the prescribed and measured velocity is sent through a simple amplification module. The result from this amplification is then sent into a control output element that couples this control output to the magnitude of the torque entity defined above. As a first design iteration the amplification is set to 5. This model now represents a closed loop control system and the modeling is completed. Fig. 5.4a shows the completed mechanism model and Fig. 5.4b shows the corresponding control model.

a)

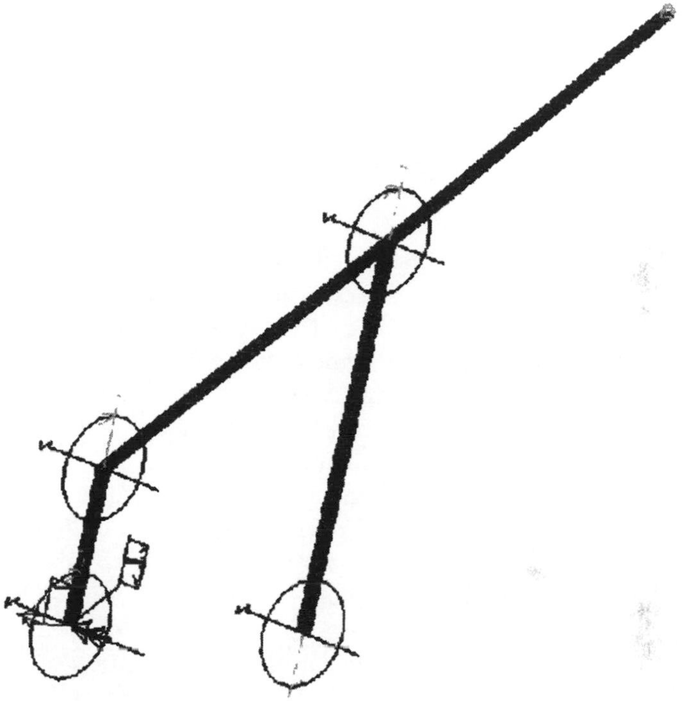

b)

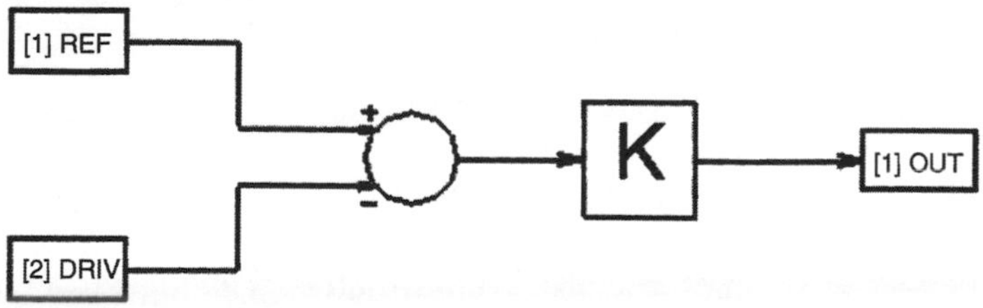

Figure 5.4: a) Mechanism simulation model. b) Control simulation model.

5.2.4 Repeated Simulations and Presentations of results

Having completed the simulation model, the solution task is entered in the FEDEM program. First the integration parameters must be initiated. As default the integration will start from zero with a time step of 0.01 seconds and an end time of 1.0 seconds. These default settings for the integration parameters could be used for an initial simulation run. Starting the simulation, first the links of the mechanism will be reduced by Component Mode Synthesis (CMS) transformation and then the simulation will start on the mechanism system model. The motion of the mechanism will be visualized on the screen during the simulation and the mechanism input link will rotate approximately $\frac{3}{4}$ of a full revolution with the simulation end time of 1.0 second.

Now the post-processing task is entered and some response curves are needed to inspect the simulation results. Having initiated the curve-plotting module, the headings for the plot are then entered. The x-axis of the diagram is defined by referring to the time entry in the menu. For defining the diagram's y-axis, "triad" is first selected from the menu and the mouse is used to pick the actual triad or if one knows the triad number this number can be entered directly. The next menu entry is used to select the global x direction for the triad and the last menu entry is used to select acceleration. The curve will be plotted using the apply menu , see Fig. 5.5. Going back to the model solution task the simulation end time is changed from 1 to 3 seconds to get a better picture of the steady-state motion. During a new simulation the input link of the mechanism now will rotate about 2 and $\frac{3}{4}$ revolution. Re-plotting the diagram for the x-acceleration of point D, see Fig. 5.6

The acceleration amplitude is about 50 m/sec^2 , that is higher than specified. A response curve is also made for the joint variable velocity at point A, see Fig. 5.7. This velocity is controled to vary about 6.28 rad/sec that is one revolution per second. The variation is well within the range 5-8 rad/sec that is specified.

The aim for the design optimization is now to reduce the x-acceleration for point D to below 40 m/sec^2 . First we will check the effect of introducing a flywheel on the motor shaft. We have already modeled the inertia for the rotor in the motor by introducing a rotational mass of 0.15 kgm^2 in the model. By increasing this mass to 1.00 kgm^2, the effect of a flywheel of inertia 0.85 kgm^2 on the motor shaft is introduced. Re-plotting the diagram for the x-acceleration of point D, see Fig. 5.8, we see that the acceleration has even increased beyond 60 m/sec^2. This effect can be explained by the

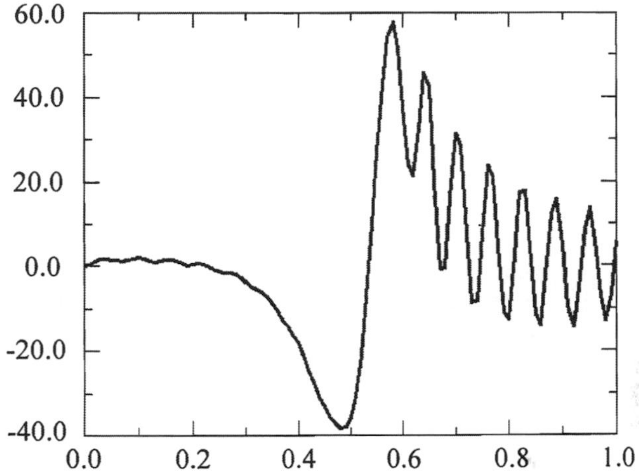

Figure 5.5: Setting up a curve plot.

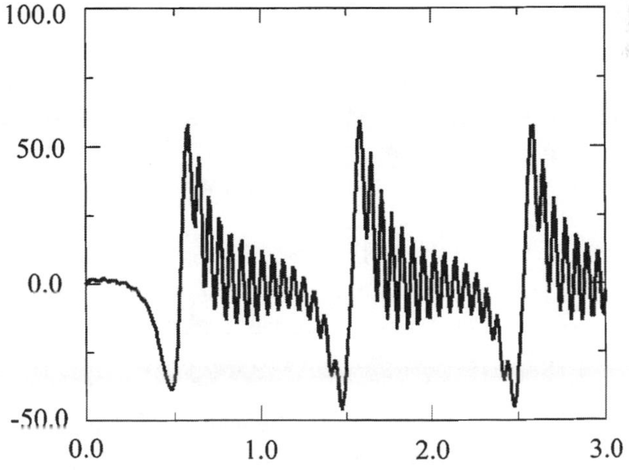

Figure 5.6: x-acceleration for point D $[m/s^2]$.

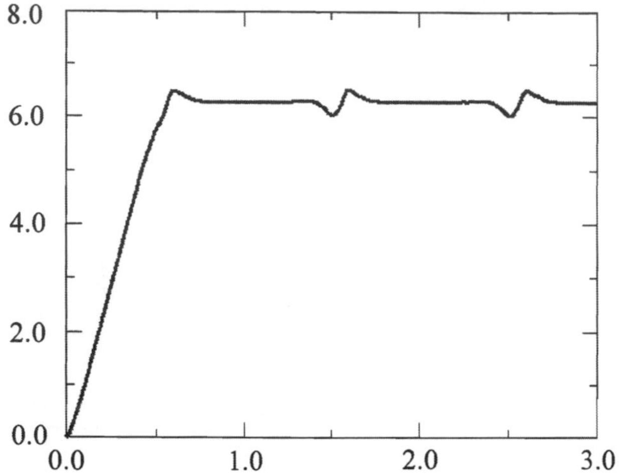

Figure 5.7: Joint variable velocity in point A $[m/s]$.

fact that a larger flywheel effect will give less variation in the input speed and have similar effect as larger control amplification. The input rotational velocity is forced to low variation and this will give an increase in the flexible vibrations of the links.

The extra flywheel effect is now removed and the control amplification is reduced from 5 to 1. The end time for the simulation is set to 5 seconds to assure that the system comes to a stable steady-state situation. The response curve in Fig. 5.9 shows the x-acceleration for point D that is within 40 m/sec^2 . Fig. 5.10 shows the input rotational velocity response that is well within the range 5 - 8 during steady-state operation. Fig. 5.11 shows the torque on the motor shaft that has a maximum of 2 Nm during start-up and this will be the design requirement when selecting the motor.

The coupler curve for point D should also be plotted. The diagram heading is entered as for the curves above, however, no axis in the diagram should now refer to time. For the diagram's x-axis the triad at point D is selected with a global x-direction and position. Also for the diagram's y-axis the triad at point D is selected, however, now the the y-direction and position are selected. The coupler curve is shown in Fig. 5.12.

The mechanism may also be analysed with respect to material stresses.

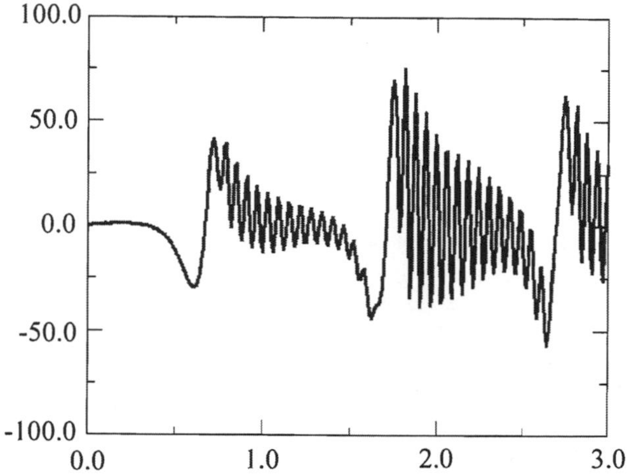

Figure 5.8: x-acceleration for point D including flywheel effect $[m/s^2]$.

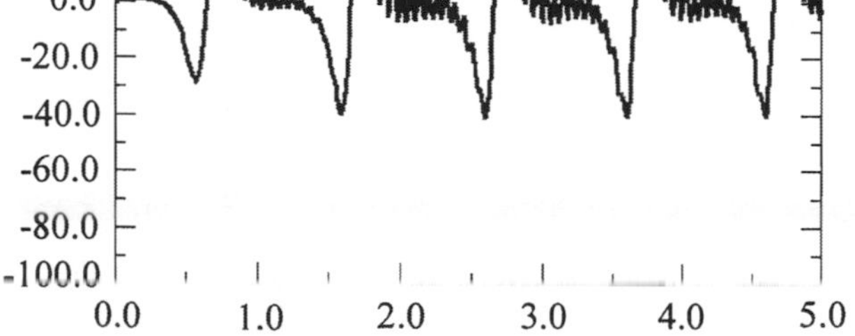

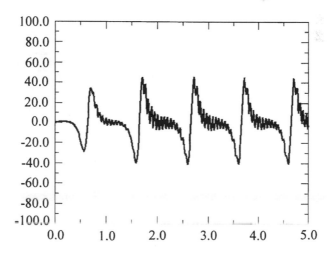

Figure 5.9: x-acceleration for point D $[m/s^2]$.

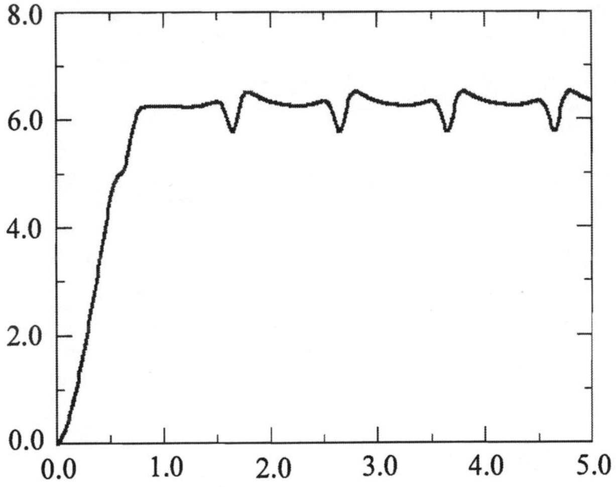

Figure 5.10: Joint variable velocity at point A $[m/s]$.

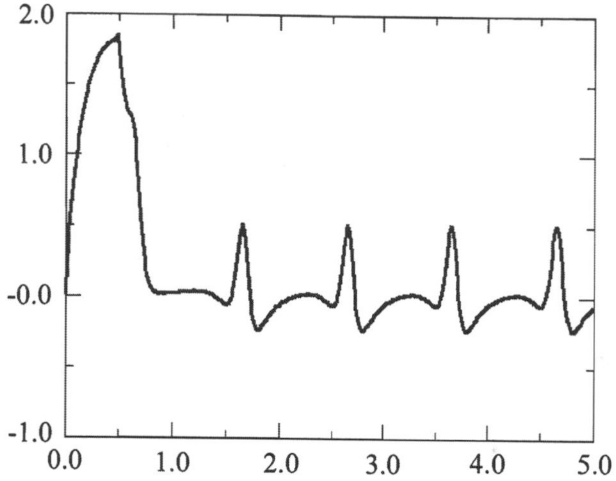

Figure 5.11: Motor shaft torque $[Nm]$.

Entering the model solution task some time steps from the simulation may be selected for stress analysis. We select the start time to be 3 seconds, the increment to 0.1 seconds and the end time to 4 seconds, that is stresses for the links will be calculated for 11 position for one revolution of the mechanism. Having selected these positions for stress analysis the stress retracking could be started. The post-processing task is then re-entered and the high performance animation module is started to animate the mechanism motion with colour stress fringes added on the links. Link deformations may also be included in the animations. A single image from the stress animation is shown in Fig. 5.13. Simulations results from a number of industrial applications using the FEDEM software can be studied on the CD that is available for free distribution by the Fedem Technology company, see the reference to the Fedem Simulation Software Demonstration CD (2001). The stress contour plots in the animation show that high stress levels are observed on the input link close to the motor shaft and in the coupler link around joint B as could be expected. Stress calculations should be carried out for more time steps during one revolution of steady-state operation of the mechanism input link to get dependable stress amplitude and frequency values for fatigue analysis. If yield stress is a problem or the start-up of the mechanism will be repeated

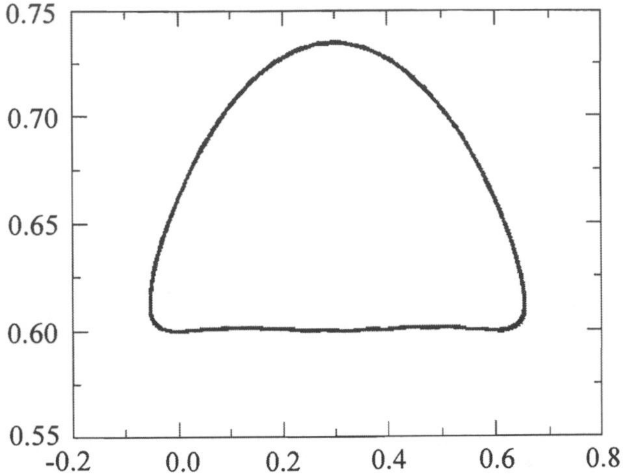

Figure 5.12: Coupler curve for point D in meters.

a large number of times and thus be a possible couse of fatigue, the stresses during transient start-up operation should also be studied in detail.

5.2.5 Discussions

The design process described in this chapter is limited to the tuning of the control system and a possible flywheel on the motor shaft for a proposed mechanism and gives the necessary data for selecting an electric motor. This simulation is also used as verification of the mechanism design both with respect to the operation and its structural integrity. No optimization with respect to the length and shapes of links were carried out.

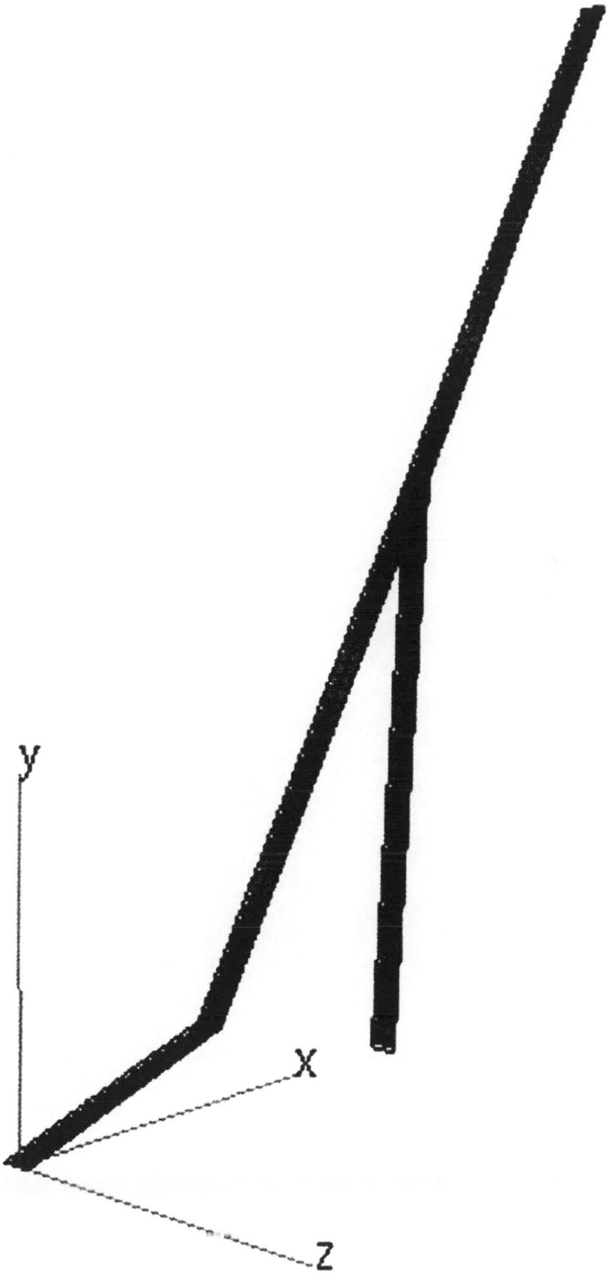

Figure 5.13: A single image from stress animation.

References

[1] Bell, K. (1987): "A General Datastructure for Linear Finite Element Analysis", SINTEF report STF71 A87026, Trondheim, Norway.

[2] Bergan P.G., Larsen P. K and Mollestad E., Svingning av konstruksjoner, Tapir 1981, Trondheim, Norway. (In Norwegian).

[3] Enright, W.H. et al. (1986): "Interpolants for Runge-Kutta Formulas", ACM Trans. on Math. Softw., vol. 12, no. 3, pp 193 - 218.

[4] "The FEDEM Theory Manual", Fedem Technology AS, Trondheim, Norway, report 1999.

[5] "The Fedem Simulation Software Demonstration CD", Fedem Technology AS (http://www.fedem.com/), Trondheim, Norway, 2001.

[6] Gear, C.W. and O. Østerby (1984): "Solving Ordinary Differential Equations with Discontinuities". ACM Trans. on Math. Softw., vol 10, no. 1, pp 23 - 44.

[7] Haugen, B., Buckling and Stability Problems for Thin Shell Structures Using High Performance Finite Elements. PhD dissertation, University of Colorado, 1994.

[8] Hildre, H.P., Machanical Design of High Performance Mechanisms. Dr.ing. (PhD) dissertation, Norwegian Institute of Technology, 1991, (now, Norwegian University of Science and Technology).

[9] Iversen, T. (1986): "Parallel, Modular Integration for Dynamic Simulation of Industrial Processes" SINTEF report STF48 F86015, Trondheim, Norway. (In Norwegian).

[10] Iversen, T. (1988): "Multidisciplinary Simulation. Method, Software Structure and Documentation for the Control part", SINTEF work note 88-63-K, Trondheim, Norway. (In Norwegian).

[11] Iversen, T., Modular Techniques in Dynamic Process Simulation. Dr.ing. (PhD) dissertation, Norwegian Institute of Technology, 1990, (now, Norwegian University of Science and Technology).

[12] Langen, I. and Sigbjörnsson R., Dynamisk analyse av konstruksjoner, Tapir 1979, Trondheim, Norway. (In Norwegian).

[13] Langen, I., Birkeland, O. and Rølvåg, T., "Simulation of Dynamic Behaviour of a FPSO Crane", Proceedings of the The 5th North Sea Offshore Crane Conference, Aberdeen, Scotland, April 26-28, 2000.

[14] Newmark, N.M. (1959): "A Method of Computation for Structural Dynamics", J. Eng. Mech. Div. ACSE, vol. 85, EM3.

[15] Rølvåg, T., Multidiscipline Dynamic Simulation of Elastic Manipulators and Space Structures. Dr.ing. (PhD) dissertation, Norwegian Institute of Technology, 1991, (now, Norwegian University of Science and Technology).

[16] Rølvåg, T. and Humphries, M., "Simulation and Verification of the GOMOS Steering Front Assembly", Proceedings of ESM98 Simulation - Past, Present and Future, Conference, June 16-19, 1998, Manchester, UK.

[17] Sivertsen, O.I., Large Displacement Finite Element Formulations of Elastic Mechanism Dynamics. Dr.ing. (PhD) dissertation, Norwegian Institute of Technology, 1981, (now, Norwegian University of Science and Technology).

[18] Åmnes, K., Visualization of Results from Mechanical Engineering Analysis. Dr.ing. (PhD) dissertation, Norwegian Institute of Technology, 1990, (now, Norwegian University of Science and Technology).

Appendix A

Corotational Geometric Stiffness

The mechanism simulated consists of links and each link consists of one or several elements. The link is regarded as a super element with a variable number of external nodes. The links are connected together through their external nodes to form a mechanism. The mechanism is simulated nonlinearly i.e. the geometry is updated throughout the analysis (the text in this section is adopted from the Fedem Theory Manual (1999)).

By adding rotational geometric stiffness (pre-stress stiffness) to the material stiffness, the stiffness matrix K will improve. If the structure is in a stressed state, we will get a more correct stiffness matrix (closer to the tangent). The convergence rate of the Newton-Raphson iteration algorithm will improve for most cases. This will also give more correct eigenfrequencies for structures in a stressed state.

A complete implementation of all stiffness terms from the corotated formulation could be carried out but we probably would not gain much profit compared with the extra CPU time needed. The material stiffness term and the rotational geometric stiffness term are the most important stiffness terms in the tangent stiffness matrix for the corotated formulation when the deformations are small.

The rotational geometric stiffness term is very sensitive to the positioning of the reference points that define the super element's local coordinate system. An automatic selection and optimization algorithm has been developed and implemented.

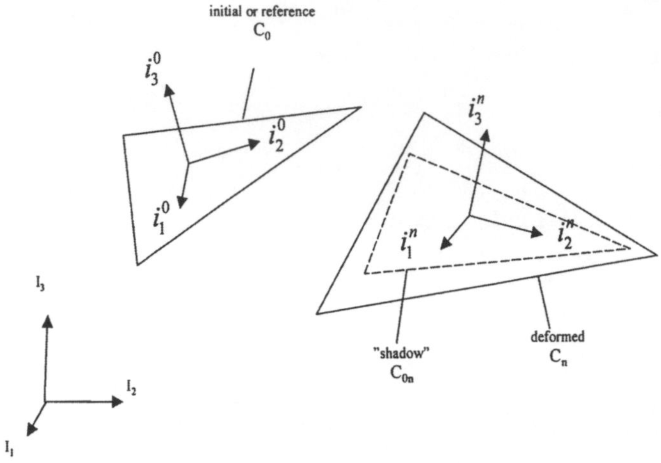

Figure A.1: Coordinate systems and configurations for an element.

Theoretical Background

In the following, super-script "$\sim$" indicates local coordinate system, no super-script indicates global coordinate system. The configurations will be denoted as following (See Fig. A.1):

C_0: initial (reference) configuration

C_{0n}: corotated (shadow) configuration

C_n: current (deformed) configuration

Since the system are solved for each load increment, the 3D rotations are considered small. Therefore, the rotations are given vector properties in the following.

Rotational Geometric Stiffness

The internal super element force vector in FEDEM:

$$\boldsymbol{f} = \boldsymbol{T}^T \boldsymbol{f} = \boldsymbol{T}^T \tilde{\boldsymbol{k}}_e \tilde{\boldsymbol{v}}_d \tag{A.1}$$

$$\boldsymbol{T}^T = \begin{bmatrix} \boldsymbol{T}_1^T & & \\ & \ddots & \\ & & \boldsymbol{T}_n^T \end{bmatrix} \tag{A.2}$$

The tangent stiffness is said to be consistent with the internal forces if it relates the variation of $\boldsymbol{f}$ with respect to variations in $\boldsymbol{v}$

$$\delta \boldsymbol{f} = \frac{\partial \boldsymbol{f}}{\partial \boldsymbol{v}} \delta \boldsymbol{v} = \boldsymbol{k} \delta \boldsymbol{v} \tag{A.3}$$

$$\delta \boldsymbol{f} = \delta \boldsymbol{T}^T \tilde{\boldsymbol{k}}_e \tilde{\boldsymbol{v}}_d + \boldsymbol{T}^T \delta \tilde{\boldsymbol{k}}_e \tilde{\boldsymbol{v}}_d + \boldsymbol{T}^T \tilde{\boldsymbol{k}}_e \delta \tilde{\boldsymbol{v}}_d = \boldsymbol{k}_{GR} \delta \boldsymbol{v} + \boldsymbol{k}_M \delta \boldsymbol{v} \tag{A.4}$$

To be consistent with Eq. (A.1) we assume:

$$\delta \tilde{\boldsymbol{v}}_d = \frac{\partial \tilde{\boldsymbol{v}}_d}{\partial \tilde{\boldsymbol{v}}} \delta \tilde{\boldsymbol{v}} \approx \delta \tilde{\boldsymbol{v}} \tag{A.5}$$

$$\delta \tilde{\boldsymbol{k}}_e = \frac{\partial \tilde{\boldsymbol{k}}_e}{\partial \tilde{\boldsymbol{v}}_d} \delta \tilde{\boldsymbol{v}}_d = \boldsymbol{0} \ \text{ since } \ \frac{\partial \tilde{\boldsymbol{k}}_e}{\partial \tilde{\boldsymbol{v}}_d} = \boldsymbol{0} \tag{A.6}$$

$$\delta \tilde{\boldsymbol{v}} = \boldsymbol{T} \delta \boldsymbol{v} \tag{A.7}$$

where

- $\boldsymbol{f}$: internal force vector
- $\boldsymbol{k}$: tangent stiffness matrix
- $\boldsymbol{k}_{GR}$: rotational geometric stiffness matrix
- $\boldsymbol{k}_M$: material stiffness matrix
- $\boldsymbol{k}_e$: linear stiffness matrix
- $\boldsymbol{v}$: total displacement vector, ordered node by node
- $\boldsymbol{v}_d$: deformational-displacement vector
- $\boldsymbol{T}_n$: transformation matrix from local to global coordinate system in C_n

We consider the rotational geometric stiffness term in Eq. (A.4) but first we need to establish the variation of the transformation matrix (see Fig. A.1).

$$\begin{aligned}
\delta \boldsymbol{T}_n &= \frac{\partial \boldsymbol{T}_n}{\partial \tilde{\boldsymbol{\omega}}_i} \delta \tilde{\boldsymbol{\omega}}_i \qquad\qquad\qquad\qquad\qquad\qquad\qquad\qquad\qquad\qquad (A.8) \\
&= \begin{bmatrix} \boldsymbol{0}^T \\ \boldsymbol{i}_3^{nT} \\ -\boldsymbol{i}_2^{nT} \end{bmatrix} \delta \tilde{\omega}_x + \begin{bmatrix} -\boldsymbol{i}_3^{nT} \\ \boldsymbol{0}^T \\ \boldsymbol{i}_1^{nT} \end{bmatrix} \delta \tilde{\omega}_y + \begin{bmatrix} \boldsymbol{i}_2^{nT} \\ -\boldsymbol{i}_1^{nT} \\ \boldsymbol{0}^T \end{bmatrix} \delta \tilde{\omega}_z \\
&= \begin{bmatrix} 0 & \delta \tilde{\omega}_z & -\delta \tilde{\omega}_y \\ -\delta \tilde{\omega}_z & 0 & \delta \tilde{\omega}_x \\ \delta \tilde{\omega}_y & -\delta \tilde{\omega}_x & 0 \end{bmatrix} \begin{bmatrix} \boldsymbol{i}_1^{nT} \\ \boldsymbol{i}_2^{nT} \\ \boldsymbol{i}_3^{nT} \end{bmatrix} \\
&= -Spin(\delta \tilde{\boldsymbol{\omega}})\, \boldsymbol{T}_n
\end{aligned}$$

$$\delta \boldsymbol{T}_n^T = \boldsymbol{T}_n^T Spin(\delta\tilde{\boldsymbol{\omega}}) \tag{A.9}$$

where $\delta\boldsymbol{\omega}_i$ is instantaneous rotation axis.

We can now establish the rotational geometric stiffness matrix:

$$\begin{aligned}
\delta \boldsymbol{T}^T \tilde{\boldsymbol{f}} &= \begin{bmatrix} \delta\boldsymbol{T}_n^T \tilde{\boldsymbol{n}}_1 \\ \delta\boldsymbol{T}_n^T \tilde{\boldsymbol{m}}_1 \\ \vdots \\ \delta\boldsymbol{T}_n^T \tilde{\boldsymbol{n}}_N \\ \delta\boldsymbol{T}_n^T \tilde{\boldsymbol{m}}_N \end{bmatrix} = \begin{bmatrix} \boldsymbol{T}_n^T Spin(\delta\tilde{\boldsymbol{\omega}}_r)\tilde{\boldsymbol{n}}_1 \\ \boldsymbol{T}_n^T Spin(\delta\tilde{\boldsymbol{\omega}}_r)\tilde{\boldsymbol{m}}_1 \\ \vdots \\ \boldsymbol{T}_n^T Spin(\delta\tilde{\boldsymbol{\omega}}_r)\tilde{\boldsymbol{n}}_N \\ \boldsymbol{T}_n^T Spin(\delta\tilde{\boldsymbol{\omega}}_r)\tilde{\boldsymbol{m}}_N \end{bmatrix} \\
&= -\boldsymbol{T}^T \begin{bmatrix} Spin(\tilde{\boldsymbol{n}}_1) \\ Spin(\tilde{\boldsymbol{m}}_1) \\ \vdots \\ Spin(\tilde{\boldsymbol{n}}_N) \\ Spin(\tilde{\boldsymbol{m}}_N) \end{bmatrix} \delta\boldsymbol{\omega}_r \qquad\qquad (A.10) \\
&= -\boldsymbol{T}^T \tilde{\boldsymbol{F}}_{nm}\delta\tilde{\boldsymbol{\omega}}_r = -\boldsymbol{T}^T \tilde{\boldsymbol{F}}_{nm}\tilde{\boldsymbol{G}}\boldsymbol{T}\delta\boldsymbol{v} = \boldsymbol{k}_{GR}\delta\boldsymbol{v}
\end{aligned}$$

$$\boldsymbol{k}_{GR} = -\boldsymbol{T}^T \tilde{\boldsymbol{F}}_{nm}\tilde{\boldsymbol{G}}\boldsymbol{T} \tag{A.11}$$

$$\tilde{\boldsymbol{k}}_{GR} = -\tilde{\boldsymbol{F}}_{nm}\tilde{\boldsymbol{G}} \tag{A.12}$$

when

$$\tilde{\boldsymbol{f}} = \begin{bmatrix} \tilde{\boldsymbol{n}}_1 \\ \tilde{\boldsymbol{m}}_1 \\ \vdots \\ \tilde{\boldsymbol{n}}_N \\ \tilde{\boldsymbol{m}}_N \end{bmatrix} \tag{A.13}$$

$$\tilde{\boldsymbol{F}}_{nm} = \begin{bmatrix} Spin(\tilde{\boldsymbol{n}}_1) \\ Spin(\tilde{\boldsymbol{m}}_1) \\ \vdots \\ Spin(\tilde{\boldsymbol{n}}_N) \\ Spin(\tilde{\boldsymbol{m}}_N) \end{bmatrix} \tag{A.14}$$

$$Spin\left(\delta\tilde{\boldsymbol{\omega}}\right)\tilde{\boldsymbol{n}}_a = -Spin\left(\tilde{\boldsymbol{n}}_a\right)\delta\tilde{\boldsymbol{\omega}} \tag{A.15}$$

$$\delta\tilde{\boldsymbol{\omega}}_r = \frac{\partial\tilde{\boldsymbol{\omega}}_r}{\partial\tilde{\boldsymbol{v}}}\delta\tilde{\boldsymbol{v}} = \tilde{\boldsymbol{G}}\delta\tilde{\boldsymbol{v}} \tag{A.16}$$

where $\delta\tilde{\boldsymbol{\omega}}_r$ is instantaneous rigid rotation axis

Our task is now to implement Eq. (A.12). The $\tilde{\boldsymbol{F}}_{nm}$ matrix is quite simple, and we will focus on the $\tilde{\boldsymbol{G}}$ matrix.

The $\tilde{G}$ Matrix

The $\tilde{\boldsymbol{G}}$ matrix relates the variation of the element rigid body rotation (the rotation of the corotated element) to the variation of the element's DOFs. The rigid body rotation of a super element is controled by the motion of three reference points ($R1, R2, R3$). These reference points define the super element's corotated coordinate system. This coordinate system is defined with origin in $R1$, x-axis along side $R1 - R2$ and the z-axis orthogonal to the $R1, R2, R3$ plane. The reference points move as they were connected through "stiff arms" to three external nodes. If the super element does not have three external nodes, several reference points will be controled by the same external node. In fact, all three reference points could be controled by the same node. In that way the super element can have a variable set of external nodes.

We will therefore establish the $\tilde{\boldsymbol{G}}$ matrix for the triangular reference element formed by the reference points, and then transform it to the super elements external nodes. The external nodes that control the motion of the reference points will later be referred to as *control nodes.* The $\tilde{\boldsymbol{G}}$ matrix for a triangular element from Eq. (A.16):

$$\delta\tilde{\boldsymbol{\omega}}_r = \frac{\partial\tilde{\boldsymbol{\omega}}_r}{\partial\tilde{\boldsymbol{v}}}\delta\tilde{\boldsymbol{v}} = \tilde{\boldsymbol{G}}\delta\tilde{\boldsymbol{v}} \tag{A.17}$$

The triangular element will not rotate while varying the rotational DOFs in $\tilde{\boldsymbol{v}}$. Therefore the rotational DOFs will not be considered. Hence, the $\delta\tilde{\boldsymbol{v}}$ vector has 9 DOFs ordered node by node. $\tilde{\boldsymbol{G}}$ can be split into contributions from each node as

$$\tilde{\boldsymbol{G}} = \begin{bmatrix} \tilde{\boldsymbol{G}}_1 & \tilde{\boldsymbol{G}}_2 & \tilde{\boldsymbol{G}}_3 \end{bmatrix} \tag{A.18}$$

$$\tilde{\boldsymbol{G}}_1 = \frac{1}{2A}\begin{bmatrix} 0 & 0 & x_{32} \\ 0 & 0 & y_{32} \\ 0 & -\frac{2A}{x_{21}} & 0 \end{bmatrix} \tag{A.19}$$

$$\tilde{\boldsymbol{G}}_2 = \frac{1}{2A}\begin{bmatrix} 0 & 0 & x_{13} \\ 0 & 0 & y_{13} \\ 0 & \frac{2A}{x_{21}} & 0 \end{bmatrix} \tag{A.20}$$

$$\tilde{\boldsymbol{G}}_3 = \frac{1}{2A}\begin{bmatrix} 0 & 0 & x_{21} \\ 0 & 0 & y_{21} \\ 0 & 0 & 0 \end{bmatrix} \tag{A.21}$$

where x, y are coordinates for the reference element's nodes ($R1, R2, R3$)

To establish the $\tilde{\boldsymbol{G}}$ matrix for the super element, we need a relation between the super element external DOFs ($\tilde{\boldsymbol{v}}_{sup}$) and the reference DOFs ($\tilde{\boldsymbol{v}}_{ref}$) (the DOFs for the reference element) (see Fig. A.2).

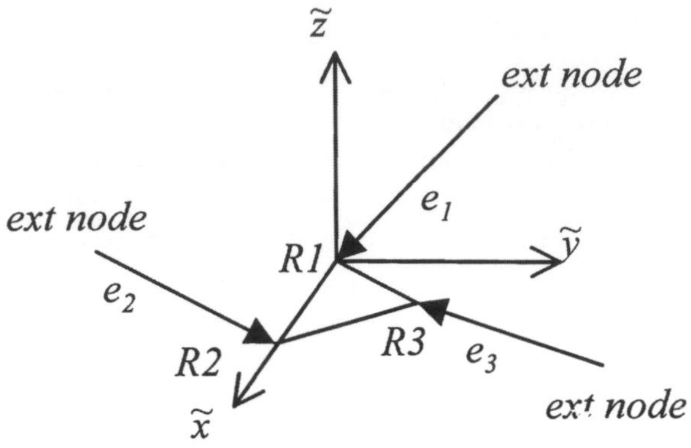

Figure A.2: The super element corotated coordinate system.

As mentioned above, the DOFs in *one* external node must be able to completely control the translations of the reference points. Therefore the stiff arms must be rigidly connected to the external node, i.e. both translational *and* rotational DOFs of the external nodes will control the translation of the reference points. Since the rotational DOFs control the movement of the corotated element, the eccentricity vectors (stiff arms) should be as short as possible to avoid the introduction of false rotations of the corotated element. The best description is gained when the reference points and the control nodes are coincident.

We must also make sure to choose the "right" control nodes, that is external nodes whose motion describes the corotated motion of the element. The rotational geometric stiffness is very sensitive upon these factors.

$\tilde{\boldsymbol{v}}_{sup}$ has dimension 6× (number of external) and $\tilde{\boldsymbol{v}}_{ref}$ has dimension 9×1, both vectors are ordered node by node.

$$\tilde{\boldsymbol{v}}_{ref} = \tilde{\boldsymbol{M}} \tilde{\boldsymbol{v}}_{sup} \tag{A.22}$$

$\tilde{\boldsymbol{M}}$ depends upon which external node moves the three reference nodes. For example, if external node 2 moves $R1$, external node 1 moves $R2$ and external node 4 moves $R3$ for a 4-noded super element, $\tilde{\boldsymbol{M}}$ becomes

$$\tilde{\boldsymbol{M}} = \begin{bmatrix} 0 & \tilde{\boldsymbol{m}}_1 & 0 & 0 \\ \tilde{\boldsymbol{m}}_2 & 0 & 0 & 0 \\ 0 & 0 & 0 & \tilde{\boldsymbol{m}}_3 \end{bmatrix} \tag{A.23}$$

when

$$\tilde{\boldsymbol{m}}_i = \begin{bmatrix} 1 & 0 & 0 & 0 & e_{iz} & -e_{iy} \\ 0 & 1 & 0 & -e_{iz} & 0 & e_{ix} \\ 0 & 0 & 1 & e_{iy} & -e_{ix} & 0 \end{bmatrix} \tag{A.24}$$

The $\tilde{\boldsymbol{G}}$ matrix for the super element ($\tilde{\boldsymbol{G}}_{sup}$) becomes for this example

$$\begin{aligned} \tilde{\boldsymbol{G}}_{sup} &= \tilde{\boldsymbol{G}}_{ref}\tilde{\boldsymbol{M}} = \begin{bmatrix} \tilde{\boldsymbol{G}}_1 & \tilde{\boldsymbol{G}}_2 & \tilde{\boldsymbol{G}}_3 \end{bmatrix} \begin{bmatrix} 0 & \tilde{\boldsymbol{m}}_1 & 0 & 0 \\ \tilde{\boldsymbol{m}}_2 & 0 & 0 & 0 \\ 0 & 0 & 0 & \tilde{\boldsymbol{m}}_3 \end{bmatrix} \\ &= \begin{bmatrix} \tilde{\boldsymbol{G}}_2\tilde{\boldsymbol{m}}_2 & \tilde{\boldsymbol{G}}_1\tilde{\boldsymbol{m}}_1 & 0 & \tilde{\boldsymbol{G}}_3\tilde{\boldsymbol{m}}_3 \end{bmatrix} \end{aligned} \tag{A.25}$$

where $\tilde{\boldsymbol{G}}_{ref}$ is the $\tilde{\boldsymbol{G}}$ matrix for the reference element as in Eq. (A.21).

Appendix B

The Cam Joint

The cam surface is defined by a set of master triads, and these nodes are ordered in increasing sequence. The z-axis is pointing in the direction of the surface tangent, and the x-axis in the direction normal to the surface. If the first and last masters are the same, the cam has a closed surface. One slave triad defines the follower. The local direction is irrelevant (the text in this section is adopted from the Fedem Theory Manual (1999)).

The cam joint is simulated by a nonlinear spring, which is connected between the cam curve and the cam follower. It is possible to connect a spring from the slave and the contact point in both x and y-directions. It is possible to add friction forces based on the forces from the x-spring.

The master triads define the surface of the cam. A new coordinate system is calculated at the contact point. This coordinate system has the z-axis in the cam surface direction and x-axis normal to the surface. The contact is enforced through nonlinear springs between masters and slaves.

For the time being the cam curve must consist of straight lines and circle arcs. Each master triad must be located exactly on the surface. Only the position of the triads is used to describe the curve. The entire curve consists of a set of curve segments, each made up of three nodes. Each curve segment is a circle arc unless the three points are identified to be on the same line. Then the curve segment is a straight line.

Although the directions of the triads are not important when drawing the curve, they are important for other purposes. The x-axis tells which side is the upside of the cam and should point in the normal direction, away from the surface. The z-axis is used to identify which curve segment the slave lies on and it should point to the next master.

It is worth noticing that a sudden change in the radii of the curve gives a discontinuity in the second derivative resulting in sudden forces in the cam.

First, we identify which curve segment the slave node is nearest to. We

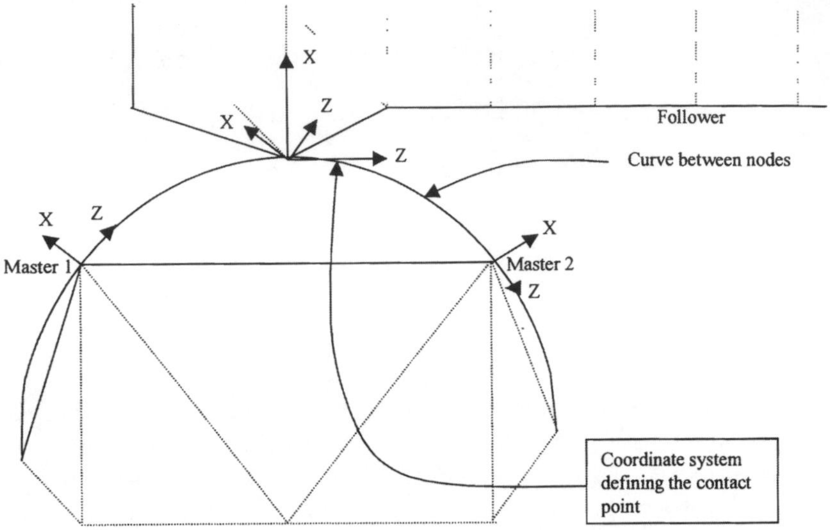

Figure B.1: Coordinate system attached to the slave node and to the master nodes for the cam joint.

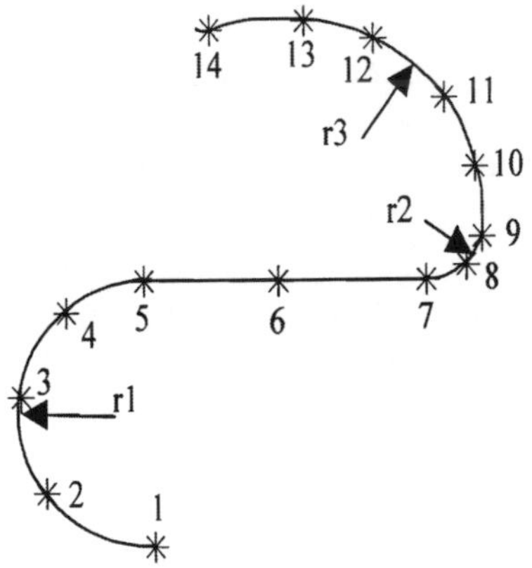

Figure B.2: Cam surface consisting of a circle arc, a straight line, a circle arc and a circle-arc. The first and last circle arcs consist of two curve segments (5 nodes). The straight line and the second circle arc consist of one curve segment.

then need to identify which point on the circle arc the spring should be connected to, being the point on the curve closest to the follower. We also need the tangent for the curve at this point. These variables are needed:

P_1: First triad on the curve segment

P_2: Second triad on the curve segment

P_3: Last triad on the curve segment

These points are given in global coordinates. A new coordinate system, $\boldsymbol{T}$, is built. $\boldsymbol{T}$ has P_1 in origo, P_2 as a point on the x-axis, and P_3 in the xy-plane on the positive y-axis. The points are then transformed to this system:

$$\begin{aligned}
P_{1local} &= \begin{bmatrix} 0 & 0 & 0 \end{bmatrix}^T \\
P_{2local} &= \boldsymbol{T}^{-1}P_2 = \begin{bmatrix} T^{-1}_{1,1}P_{2x} + T^{-1}_{1,2}P_{2y} + T^{-1}_{1,3}P_{2z} + T^{-1}_{1,4} & 0 & 0 \end{bmatrix}^T \\
P_{3local} &= \boldsymbol{T}^{-1}P_3 = \begin{bmatrix} P_{3local_x} & P_{3local_y} & 0 \end{bmatrix}^T
\end{aligned} \tag{B.1}$$

Now we have three points on the xy-plane. Ignoring the z coordinate we can find the origin of the circle. Given the equation of a circle:

$$\left(x^2 + y^2\right) + 2gz + 2fx + c = 0 \tag{B.2}$$

Where

$$Origo = OO = (O_x, O_y) = (-g, -f) \tag{B.3}$$

$$Radius = R = \sqrt{g^2 + f^2 - c} \tag{B.4}$$

$$g = -\frac{1}{2}P_{2x,local} \tag{B.5}$$

$$f = -\frac{1}{2}\frac{P_{3x,local}}{P_{3y,local}}\left(P_{3x,local} - P_{2y,local}\right) \tag{B.6}$$

$$c = 0 \tag{B.7}$$

$$R^2 = g^2 + f^2 \tag{B.8}$$

The slave is now given by a point CC, and a new point, CR, on the circle must be found. The point is found by moving CC in a straight line from the origin to CC, until it hits the circle.

$$n = \frac{R}{|CC_{local} - OO_{local}|} \tag{B.9}$$

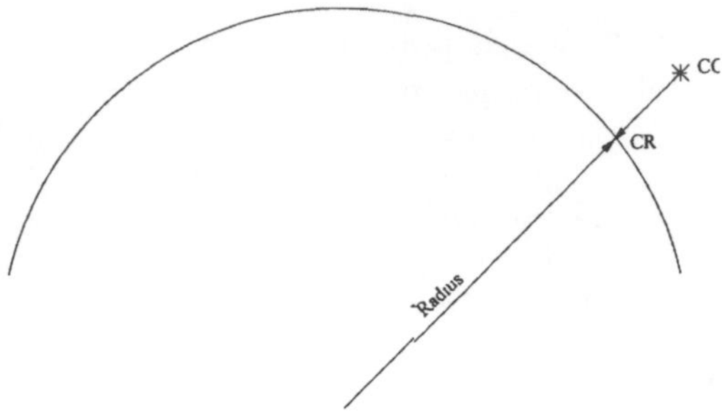

Figure B.3: The slave's closest position on the curve is found.

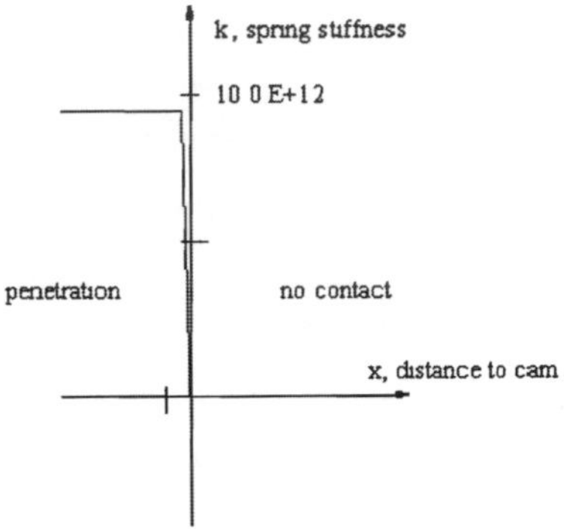

Figure B.4: The contact spring stiffness can be defined by a nonlinear spring function.

$$CR_{local} = n\,(CC_{local} - OO_{local}) + OO_{local} \tag{B.10}$$

Since we now are dealing with only two dimensions the tangent vector CRt_{local} can be found by:

$$CRt_{local} = [C_{ry}, -C_{rx}, C_{rz}]^T \tag{B.11}$$

and all points can be orientated back to their original coordinate system:

$$Pn = T \cdot Pn_{local} \tag{B.12}$$

The contact point on the cam surface is now defined. The length of the cam spring is defined as the distance between the contact point and the follower. This spring can now be given any characteristic to simulate the surface. The spring stiffness and the forces are then added to the follower and distributed between the two surrounding masters to give equilibrium. If the cam follower has an offset from the curve in the contact point's y-direction, then a moment is also added to the masters to assure moment balance. The same is also the case for the y-directional spring.

Printed and bound by CPI Group (UK) Ltd, Croydon, CR0 4YY
23/10/2024
01777667-0008

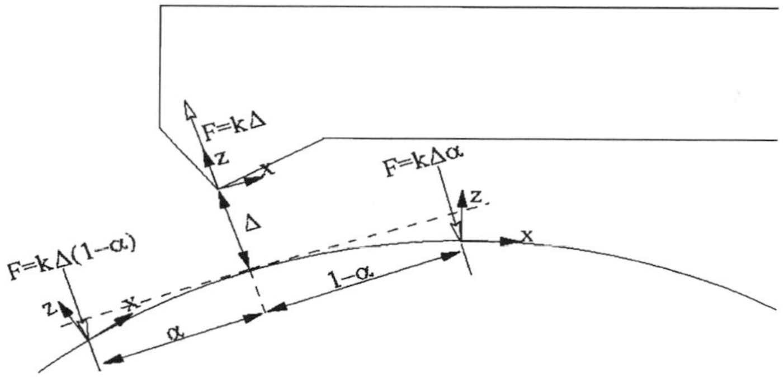

Figure B.5: A spring length is calculated between the follower and the cam surface. The stiffness and the forces are added to the masters and the slave.

ADVANCES IN ENGINEERING

1. *Nonlinear Dynamics of Compliant Offshore Structures.*
   Patrick Bar-Avi and Haym Benaroya
   1997. ISBN 90 265 1449 9

2. *Handbook of Vehicle-Road Interaction.*
   David Cebon
   1999. ISBN 90 265 1554 5

3. *Nonlinear Random Vibration. Analytical Techniques and Applications.*
   Cho W.S. To
   2000. ISBN 90 265 1637 1

4. *Virtual Testing of Mechanical Systems. Theories and Techniques.*
   Ole Ivar Sivertsen
   2001. ISBN 90 265 1811 0